WORLDS OF SCIENCECRAFT

This book is dedicated to

Sharon Anderson-Gold
Iain Banks
David Heath
Susan Leigh-Star
Anne Stingl
Helen Weiner

Worlds of ScienceCraft

New Horizons in Sociology, Philosophy, and Science Studies

SAL RESTIVO
University of Ghent, Belgium

SABRINA M. WEISS
Rochester Institute of Technology, USA

ALEXANDER I. STINGL
Drexel University, USA and University of Kassel, Germany

Routledge
Taylor & Francis Group

LONDON AND NEW YORK

First published 2014 by Ashgate Publishing

Published 2016 by Routledge
2 Park Square, Milton Park, Abingdon, Oxfordshire OX14 4RN
711 Third Avenue, New York, NY 10017, USA

First issued in paperback 2016

Routledge is an imprint of the Taylor & Francis Group, an informa business

British Library Cataloguing in Publication Data
A catalogue record for this book is available from the British Library.

The Library of Congress has cataloged the printed edition as follows:
Restivo, Sal P.
 Worlds of ScienceCraft : new horizons in sociology, philosophy, and science studies / by Sal Restivo, Sabrina M. Weiss and Alexander I. Stingl.
   pages cm
 Includes bibliographical references and index.
 ISBN 978-1-4094-4527-2 (hardback) 1. Sociology--Research. 2. Philosophy--Research.
3. Science--Research. I. Weiss, Sabrina M. II. Stingl, Alexander I. III. Title.
 HM571.R47 2014
 301.072--dc23

                                                                                    2013046906

ISBN 13: 978-1-138-27177-7 (pbk)
ISBN 13: 978-1-4094-4527-2 (hbk)

Contents

Acknowledgements

Sabrina Weiss would like to thank the following people: Ron Eglash, Deborah Blizzard, Langdon Winner, Steve Breyman, Ellen Esrock, Kelly Joyce, Ned Woodhouse, Nancy Campbell, Eben Kirksey. CSD, Michael Fitzgerald, Alex Sapadin, Liam and Kellen Potocsnak, Brandy Spani, Catherine Helle, Carly Johnson, Nika, Todd, and Sarah Spiker-Rainey, Lisa, Danny, and Zoe Fontaine-Rainen, Teresa Griffin, Anqi Fu. Snohomish and Kamiak High School Debate Teams, Steve Helman, Bill Nicolay, and Marissa Elliot. Chris Das Neves, Tyler Neal, Sean Galbraith, Zannechaos—fellow modest cyborgs. Owakeri, Lindiwe, Mimeteh, Astenya, Jolilae, Arzu, Bullzilla, and the rest of the great people at WRA. Lorn Reynolds, John Daziens, Pecos, and everyone else from ESX. Linda MacDonald-Glenn, Fred Childs, Brit Yamamoto, Allison Mead, Bill Costello, Steve Koepp, Shan Oglesby, Mrs. Spencer, Mrs. Tawlks, Carol Rindle, Mrs. Yakovich, Sister Dolores-Crosby, Cliff. My family, including my grandparents, aunts, uncles, and cousins around the world; Dan, my brother, for being the best big little brother a person could have, and Alexander I. Stingl for being my partner and equal. And most of all, to my parents for both challenging and supporting me through the many adventures of my life.

Alexander I. Stingl has profited from audiences and co-presenters of talks between 2007 and 2013. He would also like to thank his students and TAs (Tutoren) at Leuphana University Lüneburg who engaged and discussed literatures and knowledges with him and happily challenged his conventional views and his provocations alike over the past few years. A particular thanks to Hans Bakker for reminding him of the importance of semiotics and Max Weber, and also for serving as the "image of a woman" at SSSI in NYC in 2013. He wants to thank Kelly Joyce, Ali Kenner, Linnda Caporael, Ron Eglash, Gareth Edel, Jon Gluck, Matt Hoffman, Liz McGoey, Deborah Blizzard, Sebastian Deterling, Ellen Langer, Michael Dellwing, Elisabeth Beck-Gernsheim, Andreas Jürgens, Wendel Burkhardt, Don Levine, Fritz Breithaupt, Volker Wegener, Gayle Goldstick, Joana Mangiuca, Albero Asencio, Ferdie Wagner, David Lindner, Kathrin Schwalbe, Pete Loranger, Laura Mauldin, Jonah Friedman, Justin Grosslight, Armin Monsorno, Bernhard Tilg, Donald Levine, Lazarino Antini, Birgit Dörle, and David Heath for discussions, comments, and/or advice that were relevant to this book. He wants to emphasize he is, as always, indebted to the deterritorializing and detemporalizing dialogue with Sabrina M. Weiss. And, finally: Für Sepp und Mu, danke für alles. Ich drück euch.

Sal Restivo thanks Randall Collins, Leslie Brothers, Linnda Caporael, and Julia Loughlin for provocations that have helped him to better grasp the social nature of mathematics, brain, and mind.

Prologue

Sal Restivo and Sabrina M. Weiss[1]

There are always dangers when we hook our wagon to a star, to an icon of our field or an iconic idea. Notably dangerous icons include Einstein in physics, Foucault in history, Durkheim in sociology, Plato in ancient philosophy, Hegel and Kant in modern philosophy, and Marx in political economy. Dangers also accrue to iconic ideas such as relativity, entropy, complementarity, uncertainty (Heisenberg or Gödel), structuralism, social constructionism, and historical materialism. The dangers are multiple: we must confront the challenges of translation in some cases, the internal contradictions that inevitably crop up in a lifetime of writing, and more often the conflicting interpretations among authors themselves and their readers (T.S. Kuhn being a notable exemplar); and we must be alert to the threat of anachronisms and Whig histories. Finally, we must be cautious about falling prey to hero worship. This danger exists wherever we encounter idolatry, whether in the case of iconic figures, core concepts, ideas and themata, or through disciplinary perspectives.

Sal Restivo, Ghent, Belgium, 2012

Inspirations

The inspiration for the title of this work, "Worlds of *ScienceCraft*," came from the popular massively-multiplayer online roleplaying game (MMORPG), *World of Warcraft*. This connection arose out of a series of discussions between Restivo and Weiss and only later was crafted as an homage to this title.

The game raised the question about the nature and quality of social interaction. Are online social interactions functionally equivalent to offline face-to-face interactions? This question raised deeply intersectional juxtapositions between disciplines (sociology vs. philosophy), generations (old vs. young), cultures (western and eastern), and gender (masculine, feminine, and queer). For the sociologist, humans are always, already, and everywhere social and the face-to-face aspect of our species defines our humanity. But for a bioethicist philosopher, there is undeniable value in recognizing the reality of interactions, regardless of medium or participant, because all have emotional and conceptual content and impact.

The discussions between Restivo and Weiss eventually led to a recognition that just as, to paraphrase William Blake, individuals could be multiples so could

---

1 The "Inspirations" section was written by Weiss; the rest of this prologue represents an introduction in Restivo's voice.

social interactions be multiples. If face-to-face is ideal, then what of people with disabilities who cannot interact in the same ways because they can't hear, or see, or smell? What if someone is disfigured and their face cannot convey natural expressions? Are we to say that these people are not experiencing "true" social interactions and therefore are not truly human? In recognizing the interconnectedness of the variety of human (and nonhuman) interactions, Weiss and Restivo came to embrace a tenuous but more generous worldview that valued the importance of being to knowledge-crafting (the core of what science, small "s," is) and our understandings of the world to our being. At the end of the day, different kinds of social interactions in different settings will produce different human beings. Human–human, human–machine (social or sociable robot), human–computer interface, human–animal all have their specific qualities and potentials for producing a certain type of human being, a certain type of interactional dynamic, and a certain type of emotional life. There may be reasons for preferring one of these relationships over the rest but once we are in the midst of the multiples we must be prepared to give reasons for our choice and defend it comparatively and contextually.

So now we could construct a Whig history of this homage title and point to the similarities between objectivity communities and factions (Alliance or Horde), where everyone in the same group speaks the same language, shares the same view of the world, and often collide hostilely with those of other communities/ factions. Or we could point to the parallel multiplicities of being where many servers exist (and indeed, regions of servers based on spatial geographies of Europe, North America, Asia) and the same types of events happen on each server, with mirroring social structures and activities, but with interchangeable actors and slightly different demographics based on structural rules. These rules are enacted in *Warcraft* and similar games through server types, such as "Player-vs-Player" (PvP) where player characters are allowed or even encouraged to fight against each other, or "Roleplaying" (RP) where players are expected to act "in character" at all times except in out-of-character (OOC) discussions—this means that you are not allowed to speak about "real world" events (like work, your commute, or an anticipated chicken dinner) so that the fantasy atmosphere of the game may be preserved. Just as we see in the offline world, these sorts of structural rules both select for and cultivate different types of actors and interactions between actors. We could also discuss the knowledge creation of the elites (see Chapter 1 in this volume) in the hyper-rational "theorycrafting" that some "hardcore" players engage in to calculate in minute detail the numbers involved in their gameplay to maximize the abilities of their characters; one of the most popular websites is self-titled "Elitist Jerks" to recognize the exclusive nature of this activity and regularly posts complex equations and calculating spreadsheets that back-engineer in-game calculating processes. We even might offer a commentary on the addictive and seductive nature of online gaming as a way for people to feel a sense of control over their world, just as we do through technoscientific regimes, even if there is as much variety within the population of WoW players in terms of motivations and proficiency level as we see in any other social activity. Weiss would argue that it is the very same social that drives us to MMOs and social networking that we indulged when gossiping in little hamlets of a couple dozen people centuries ago; all that has really changed is the quantity and style, much as our greater access to processed food combined with metabolisms geared toward preventing starvation have contributed to a growing obesity epidemic.

But those are merely reflections (rationalizations) after the fact. This book was born in a contentious dialogue between two scholars who stubbornly argued their perspectives and who decided to seek coexistence through this book. Eventually we will introduce the concept of Thirds when we move the unit of social life from the dyad to the triad. In the process of writing this book we acquired our own Third in the form of Stingl: in so doing we have managed to performatively enact the shift from dyadic to triadic interactions, and we are the richer for it.

This book is the result of this continuous process of becoming and knowing and intra-acting, and it is also a beginning. Welcome to these Worlds of *ScienceCraft*.

Crafting and Recrafting *ScienceCraft*

In her book, *Crafting Science* (1996: 3), Joan Fujimura describes the representational work of biological scientists working on the causal properties of the proto-oncogene that produced cancerous cells as follows: "... researchers crafted and recrafted the proto-oncogene theory as well as links to problems of many different scientific worlds. Bridges were built between existing lines of research, which extended and transformed them into new lines of research." This is one of the provocations for our title. The other provocation is the massively multiplayer online role-playing game (MMORPG) *World of Warcraft*, which signals our concern with drawing out the implications of gaming for re-thinking the nature of the self in relation to others and to science and technology. One of us, Weiss, is a gamer who has played *World of Warcraft* and developed sociological and philosophical insights that we play out in this book.

We begin where Restivo and Weiss began, in conflict over the value of philosophy and the strengths and weaknesses of sociology. Restivo is well known as a strong proponent of the sociological cogito (introduced by Randall Collins), C. Wright Mills' the sociological imagination, and his own versions of sociological materialism, social constructionism, and a strong form of sociological structuralism. He is also known as a critic of philosophy, having called for the "end of epistemology" and even the end of philosophy. And, in what is a paradox in appearance only, he has influenced and been associated with philosophers of mathematics noted for their turn to practice. Stingl's contributions here reflect his strong but not uncritical affection for and profound knowledge of philosophy.

Against the Philosophers and For a Relevance Renaissance in Philosophy
(Restivo's voice)

In 1988 the celebrated French anthropologist Claude Levi-Strauss (1988) in *De près de loin* asked himself: "Do you think there is a place for philosophy in today's world?" He said that there was but only if it based itself on contemporary science. Science, he wrote, has given us new views of life and our universe and at the same time revolutionized the rules of reason.

The Nobel physicist Richard Feynman complained that philosophy was just a lot of "meaningless chewing around." He had in mind in particular Spinoza. He claimed that if you inverted Spinoza's claims you wouldn't be able to tell which of the versions was true. He also complained that the ways in which philosophers describe science do not match science as he knows it. He was once asked in a Princeton University seminar if an electron was "an essential object." In order to try to clarify the meaning of the word "essential," he asked if a brick was an essential object. No two philosophers could agree on an answer to this question and the seminar ended in confusion. Feynman also claimed that philosophers ask "pre-decided questions."

Another Nobel physicist, Steven Weinberg, complains that philosophy is of little or no value for physics. Like Feynman, he doesn't see that philosophers can offer scientists anything useful about how to do science or what they are likely to discover. The views of Feynman and Weinberg are echoed by philosopher George Gale, who claims that only the smallest number of working scientists could find anything interesting in the "arcane" and "scholastic" discussions carried out by philosophers of science, One is reminded here of Wittgenstein's remark that it is very unlikely that any scientist or mathematician will find anything in his work that might influence the way s/he works. Except for Wittgenstein and Feyerabend, Weinberg finds that philosophical insights pale by comparison with the achievements of physicists and mathematicians. He describes philosophical writings as a murky, impenetrable jargon of thinkers who confuse obscurity with profundity.

The author Tom Wolfe called Edward O. Wilson the "new Darwin." Wilson, the father of sociobiology, finds in Carnap and the logical positivists an inability to make some crucial distinctions (for example, between fact and concept, empirical generalization and mathematical truth, theory and speculation) that would allow them to understand and articulate the difference between science and non-science. In his book, *Consilience* (1998), Wilson has much to say about the poverty of philosophical understanding and philosophical achievements.

In general, the anti-philosophers claim that philosophy as a method of inquiry is inadequate; it relies on insights, intuitions, and thought experiments instead of on scientific facts. Thought experiments, incidentally, should be (considering their use in physics in particular) in principle testable and not rocking chair speculations. The methodology of the philosophers relies on chains of "reasons" which are not subject to the same constraints mathematicians put on their logical chains. The methodology has no built in, commonly agreed upon, means for correcting errors. Philosophical discourse often has no stop signs or dead ends and carries on without any criteria for closure. It is the philosopher Richard Rorty's imperative to "keep the conversation going" in extremis.

Other typical criticisms are that philosophical knowledge is not trustworthy, that there are no agreed-on criteria for "progress" or advancement, that concepts don't have the more robust clarity of scientific concepts. Science studies research has made the supposed clarity of scientific concepts problematic but not in a way that is detrimental to science as a knowledge generating activity and process. Taking all these criticisms and observations into account, philosophy does not appear to be a well-defined academic subject but rather a variety of subjects in a variety of academic communities. Relying on the ideas of Thomas Kuhn and Imre Lakatos (themselves subject to criticism as philosophers), or otherwise philosophies that counter the criticisms, one can say that philosophy has no paradigms and no progressive research programmes.

Critics of the utility of philosophy, notably scientists, assume that for philosophy to be useful it must be narrowly useful for what they do professionally and/or that it must follow the canons of their own field when it comes to methods and theories. Scientists have asked similar questions about science studies. They want to know how these fields can help them do their work, or help them do it better. These sorts of criticisms are short-sighted given the variety of goals philosophers pursue. On the other hand, such critics are more on the mark when philosophers make claims about how researchers and thinkers should go about their tasks. Critics often ignore the ways in which philosophers can be the handmaidens, apologists, and ideologues of the fields they study, as has often been the case in the philosophy of science. This doesn't mean that philosophy isn't vulnerable to criticisms based on disciplinary assumptions that cut across all the modes of traditional philosophical discourse, as I will show further on.

Philosophy has been variously defined as the study of the foundations of thought and action, as a second-order "Queen of the Sciences" concerned with the logic and general reasoning procedures of the variety of intellectual arts and sciences (the arbiter of their logic and reasoning). Philosophy can be a "philosophy of ..." (e.g., of science, of logic, of physics) or a theory (e.g., Hintikka's "independence friendly logic"), "talk about talk" (Wittgenstein), the study and defense of a worldview, inquiry into wisdom, justice, virtue, morals and other aspects of "the good life" and "the meaning of life," an invitation to thinking about life, and an inquiry into the general properties of abstract ideas and concepts (facts, universals, beauty etc.). And philosophers sometimes see themselves as identifying and removing the impediments to knowledge.

Philosophers like Daniel Dennett (2006) see themselves as specialists in asking questions, not answering them. They specialize in cleaning up logical messes, including their own. John Locke (1690), in his "Epistle to the Reader" (*Essay Concerning Human Understanding*) defends this position in these terms: "... it is ambition enough to be employed as an under-labourer in clearing the ground a little, and removing some of the rubbish that lies in the way of knowledge ...."

What is the future of philosophy in the context of the preceding criticisms and definitions? One possibility is that the various branches of philosophy will be absorbed by already-existing disciplines or new interdisciplinary faculties (e.g., evolutionary biology, literature, etc. or a new Faculty of Consilience (after E.O. Wilson)). The upshot of all this is that some of us are no longer amused by speculative analyses and commentaries and have called for an end to philosophy altogether or a renaissance to relevance.

When philosophy goes empirical it goes scientific; and then it is physics, biology, chemistry, sociology, or whatever. I have had my licks concerning the end of epistemology (pushing Rorty over the edge of that abyss when it turned out at the end of the day that he couldn't go through with it and was going after all to save the jobs of the epistemologists). I have indeed called for an end to philosophy to the extent that it is the educated person's theology, and to the extent that the philosopher claims—as a living breathing philosopher I know claims—that s/he can literally sit in an armchair and come up with real-world solutions to real-world problems by using the once and only irrefutable logic that is the only tool in her/his tool kit. Weiss and some of my colleagues in philosophy have not changed my mind about SUCH philosophers, and of course I have always been at one with Nietzsche on the need for "philosophers of the dangerous," philosophy with a hammer. I might want to give some consideration to Wittgenstein's take on philosophy and "... exclude from our discussion questions which are answered by experience. Philosophical problems are not solved by experience, for what we talk about in philosophy are not facts but things for which facts are useful" (from the early 1930s lectures, Ambrose 1979). Where then do these facts that are useful come from? And is this philosophy or some hybrid philosophical anthropology (see especially his remarks on the nature of mathematics and on pain)? Perhaps the philosopher is an intellectual gadfly, subversive, the best trained and most widely travelled interrogator, the role model for the "sage," the "wise one." Perhaps I am, after all, as many of my friends and enemies have claimed, at the end of the day a philosopher. Why do they say that about me?

I am not the first or only thinker to find philosophy useless. Albert Einstein, cited here for his iconic status and not so much for his intellectual acumen, said that all of philosophy seems to be written in honey. At first, it looks wonderful, but on a second look it is gone and only mush remains (in Rosenthal-Schneider 1980: 90). Al-Ghazali (1058–111), arguably the most influential Muslim philosopher after Mohammad, wrote a treatise on "The Incoherence of the Philosophers." This is an impressive title, but it does not encompass all of philosophy, only metaphysics. He claimed that the metaphysicians did not rely on the same canons of logic as philosophers associated with the various branches of science such as physics and chemistry. The problem with Al-Ghazali's refutation of Islamic neo-Platonism (grounded in the Hellenistic tradition) is that it was carried out in defense of faith in opposition to science.

In identifying with the spirit of Durkheim, I have associated myself with Durkheim's defense of sociology in the face of philosophical and psychological adversaries. The explanation for what can appear to be disciplinary imperialism is that sociology has some more growing to do; it is still building a robust disciplinary profile. For this reason, even as we move necessarily into interdisciplinary arenas we must practice a little disciplinary imperialism. The Copernican sociology revolution still needs to undergo its Newtonian and Einsteinian revolutions. Thus there will be for some time a tension between the Durkheimian imperative and the interdisciplinary imperative. Continually ceding territory and concepts to neuro- and life scientists, to philosophers, and even to physicists and cosmologists is not what we should mean by "interdisciplinary."

In the sociology of science today this plays out in the "science wars" and within the field of science studies where Bruno Latour has recently begun to defend his actor network theory and his view of sociology by drawing on the sociology of Durkheim's contemporary Gabriel Tarde. Tarde is much more psychological than he is sociological, but this is why the individualistic Latour is drawn to him as he transforms more and more into a metaphysician. I am at odds with Latour on the philosophy issue, with Latour defending a veritable metaphysics of politics and science and me claiming that such an approach is necessarily out of touch with reality.

My objection to philosophy probably needs to be tempered. I have been strongly influenced by Nietzsche and Wittgenstein, for example. My opposition to philosophy is at one with the mathematicians Courant and Robbins when they write that we have to look to active experience, not to philosophy, to understand mathematics. In their classic text of 1906, they challenged the idea of mathematics as nothing more than a set of consistent conclusions and postulates produced by the "free will" of mathematicians. Here then is the crux of the issue: If philosophy is going to be useful it must turn to practice, to active experience. The problem is that if philosophy does this for physics, it becomes physics; if it does this for chemistry, it becomes chemistry; if it does it for any science, it becomes that science. Or it becomes the study of active experience and practice in that science, that is, it becomes social science of science. There is plenty of evidence for this turn of events. The triumphs of science have raised the question of why we still need philosophy. This is the root of the conflict between sociology and philosophy.

What kind of knowledge can philosophers lay claim to? What, at the end of the day, do philosophers know? The philosopher Gary Gutting has explicitly addressed this question in his book, *What Philosophers Know* (2009). He writes from the perspective of analytic philosophy. Based on a set of case studies (including, for example, Quine's "Two Dogmas," Plantinga's philosophy of religion, and Rorty's studies of philosophy, truth, and objectivity) Gutting concludes that philosophers have garnered "highly significant knowledge" about convictions and "first order knowledge" about fundamental questions. Philosophers, he claims, are specialists in making distinctions. Philosophers themselves, Gutting claims, should reflect more on what we learn by refining basic distinctions, and on why we continue to disagree about fundamental problems.

Gutting concludes by illustrating how what philosophers know can illuminate the nature of religious belief for non-philosophers. On the question of whether arguments for or against religious beliefs are cogent or not, philosophers offer a decisive answer; they have not and cannot prove or disprove the existence of God, they can only demonstrate that none of the standard popular arguments for or against the existence of God are adequate or compelling.

In general, philosophers' knowledge claims reflect their roles as skeptics, apologists, and experts. In the first place, their knowledge claims are based on their skills in the domain of conceptual and linguistic distinctions. Secondly, they are not masters of some unique set of modes of thinking, reasoning, and logic. The modes they rely on are widely available. Gutting acknowledges these last two points.

My reply to Gutting is that empirical findings and grounds for truths are always fallible, corrigible, and tentative; that scientists are by virtue of their profession expert skeptics and concerned in their everyday routines with conceptual and linguistic distinctions; and that empirical results can be subject to critical conceptual and linguistic analysis without being trumped by such analyses. Gutting's philosophy of God is useless once we have a scientific proof (which can be informal or a consilience of evidences as opposed to formal or otherwise rigorous). The utilities of philosophy tend to become subject to and absorbed by the unfolding of the sciences and their expansion into ever-new realms of experience.

I have at times sounded like someone who wanted to shoot all the philosophers, or at least erase the very idea of philosophy. I did, after all, open a lecture in Uppsala Sweden in 1991 by saying to the philosophers, "This town [science studies] isn't big enough for both of us." Having been persuaded by his co-authors that there is something important about philosophy, I now ask how we should proceed as sociologists in a way that embraces philosophy?

Philosophy, indeed, can be viewed as a general Platonism and equally detrimental in its classical forms and agendas to our efforts to ground mathematics (as well as science and logic) in social life. It is to philosophy in its more recent turn to practice that we must look if we are going to salvage philosophy as a credible intellectual activity. Gutting remains too committed to the classical "purist" methodologies of conceptual, linguistic, and logical analysis. The strengths of philosophy as a guide to reasoned speculation, thought experiments, and ethical stands will be heightened by the turn to practice and the empirical arena. This is how we can save philosophy as an analytical tool for understanding science and technology. And here is where philosophy is drawn into the interdisciplinary boundary breaking modalities of late twentieth and early twenty first century intellectual life. We are witnessing the demise of the traditional disciplines as we have known them for more than 100 years. As the new inter-disciplines come to the fore, it begins to appear that we are witnessing the emergence of a second-generation natural philosophy. Our goal of constructing a philosophical sociology of science is designed to reinforce the movement toward a new natural philosophy. In order to dismiss old intellectual baggage perhaps we can begin to think in terms of "natural inquiry" or "neo-natural philosophy."

Consider Michel Serres's conception of philosophy, which Weiss endorses. The philosopher defends and protects the possible, s/he shepherds the possibles, s/he is a horticulturalist in the garden of possibles, s/he is the guardian of the old-growth forest of possibles. To be a philosopher is to be the shepherd, guardian, and sower of multiplicities. For Serres, the philosopher yields the ground of the rational, essence, and truth to the functionaries of rationality and truth, the politicians and scientists. The philosopher roams the margins of politics and science where possibles are free to roam and where hope and freedom reside. Hers/his is the realm of the unforeseeable, the fragile, of instability, of innovation and multiplication, where the contingencies and confluences that close ramifications and bifurcations meet h/er resistance. S/he rules in the negentropics.

The problem here is twofold. First of all, Serres's description of the philosopher is compatible with Bohm's notion of the scientist and Nietzsche's notion of the thinker (beyond science and philosophy). Second, there is a dark side to philosophy, as there is to science; indeed if we pay attention to Nietzsche it is the dark side of thinking that makes it invaluable (but there are dark sides and there are darker sides). A philosophical science studies, then, cannot embrace a philosophy without stop signs, but it can embrace a philosophy that is at once Serres-ian, Bohmian, and Nietzschean. That is the threshold my collaborators and I plan to cross in what follows.

Moving Forward through ScienceCraft

Our collaboration is conditional. We have included previously published materials that originally appeared under our names as sole authors. Those materials have been reworked collaboratively to different degrees, but we have included the name(s) of primary authors for each chapter to account for some differences of voice. We all accept our collaborative responsibilities for each chapter but at the same time retain sole responsibility for ideas and concepts previously published. Our objective has been to allow our convergences to give birth to a next iteration of technoscience that we call *ScienceCraft*.

This book emerges at the intersection of three different biographies, three different intellectual paths, three different educational and training regimens, and two generational trajectories. If you come to this book looking for consistency, ultimate answers, coherence, and lucidly constructed orders you will be disappointed. If you understand that our intersection is also the intersection of a postmodern moment, an inflection point, a cusp characterized by a movement from old to new cultural and epistemic regimes you will be better prepared for the journey you are about to embark on. In this liminal age, we have mustered all of the resources we have at our disposal to get a glimpse of what lies ahead in the post-liminal age. Our focus is on the field cum discipline of science and technology studies but we cast a broader net across the landscape of science, technology, and culture. Having outrun the utility of our classical and modern categories and classifications, we are as a global ecumene and a species trying to escape the *cul de sac* we have cornered ourselves in. Whether or not we are in a better position than any other members of our species, biologically, culturally, or epistemically to solve this problem we are at the very least forced by our training and professions to engage the problem. We are mindful that as species adapt to their environments, they lose their adaptive potentials—thus the well-known fact that nearly all the species that have emerged on earth are now extinct. Can humans find a way to adapt without losing adaptive potentials?

This book is the story of three thinkers in search of a way out of the liminal trap, trying to find our way to some light at the end of the postmodern tunnel. Each of our chapters represents a picking away at the old regime and a search for the elements of a new regime. This amounts to constructing a new adaptive strategy, or multiple more or less integrated adaptive strategies. We cannot say with certainty what the new regimes will look like, but we know that unless we unhinge the old regimes no new regimes will emerge and our *cul de sac* will become the burial grounds of our civilizations and our species. We look to certain key civilizational and cultural niches for the clues and resources we will need to survey the next part of our human journey. Each of our chapters reflects one of these key niches. The barriers we must cross to get to the post-liminal regimes are the purity of the sciences and mathematics, the idea of progress, the mind as the secular version of the soul, "brain in a vat" perspectives, the classical uninvestigated ideas of bodies and machines, the very idea of life itself, and the nature of health and healing. These stand out for us opportunistically by virtue of our own researches, but we are well aware of other barriers that stand outside of our immediate realms of experience and competence, and we draw attention to these too without attempting to be exhaustive. We make every effort to go beyond criticism and transform our vision of the liminal present on the courts of science, technology, health, and public policy. All we claim at the end of the day is an unfinished symphony of criticism, theory, and policy for a viable human future.

Chapter 1
Mathematics, Society, and Social Change[1]

Sal Restivo

The sociologist of mathematics is concerned in the first place with what mathematics is. Traditionally, mathematics has been the arbiter of the limits of the sociology of knowledge. For most of its history, mathematics has been considered a form of knowledge that had no roots in society, culture, or history. The idea that mathematics has a transcendental nature is an inheritance of the Pythagoreans and Platonists. Plato imagined a transcendental realm populated by "Forms" or "Ideals." The purity of the Forms was manifested in the ideas, concepts, and objects we experience in our everyday world. The Platonist tradition has long held sway in the history and philosophy of mathematics.

This over-simplifies Plato but is consistent with a long tradition in the history and philosophy of mathematics. Alfred North Whitehead (1929: 63), for example, famously called the history of philosophy "footnotes to Plato" and added his own footnote on *eternal objects*. It has furthermore been suggested that Plato is himself but a footnote to Parmenides.

Consider, for example, the case of non-Euclidean geometries (NEGs). Their development in the late 1800s has been described as remarkable in three respects. First, the claim is made that they emerged in three different milieux, Göttingen,[2] Budapest, and Kazan, independently of one another. The second claim is that Budapest and Kazan were not central arenas in world mathematics. Third, the development is portrayed as startling given the simultaneity of the inventions by Riemann, Lobachevsky, and J. Bolyai. These claims express a Platonic, transcendental view of mathematics. This perspective on mathematics showed up in the sociology of knowledge and science developed from the 1920s onwards by Karl Mannheim, Robert K. Merton and their followers. Mannheim wrote in 1936 that $2 + 2 = 4$ exists outside of history; and Merton's sociology of science embraced the social system of science but exempted scientific knowledge on the grounds that it was independent of society and culture. Merton was ambivalent about the claim that scientific knowledge existed outside of society, culture, and history.

---

1 I am indebted to Randall Collins for collaborations and conversations over many years that led to the perspectives, factual materials, and theories reported in this chapter.

2 A number of German universities, in particular Göttingen, were hubs for students from abroad—in particular from America—because their interdisciplinarity-oriented laboratories were extremely well-equipped and standardized, unlike French laboratories (Canguilhem 2000: 104).

For example, when the historian G.N. Clark criticized Boris Hessen's historical materialism of Newton's *Principia*, Merton (1967: 661–3) stepped in with an eminently sociological reply. Clark defended the "purity" of Newton's scientific motives. Merton argued in opposition that individual motivations are subservient to social structures, and he made his case to support Hessen's argument. He shows an appreciation for the social construction of scientific knowledge that belies his opposition to the social constructionism arguments that marked the development of the science studies movement.[3]

A review of the history of NEGs from a sociological perspective reveals that that history begins with the earliest critics of Euclid's geometry. In the following centuries, mathematicians including Saccheri, Lambert, Klügel, Legendre, and others contributed to the emergence and crystallization of NEGs in the works of Lobachevsky (1793–1856), Riemann (1826–1866), and J. Bolyai (1802–1860). The initial provocation identified by Euclid's critics was that the fifth postulate, the *parallels postulate*, failed to meet the axiomatic self-evidence test of the first four postulates. The fifth postulate did not follow from the first four postulates. The crucial factor for the sociologist examining this case is that contrary to the views of the purists, Riemann, Lobachevsky, and J. Bolyai were part of a network at the center of which we find Gauss, (1777–1855). Gauss was working on NEGS as early as the late 1700s. Over the course of the development of NEGS, Gauss was both a facilitator of and an obstruction to the spread of the new ideas.[4]

The NEG network in its broad outlines includes J. Bolyai, the son of one of Gauss' friends, W. Bolyai. Gauss and W. Bolyai were at the University of Göttingen where the parallels postulate was "in the air." Kastner had lectured on the topic, and dissertations written on NEGs. Gauss was Riemann's dissertation supervisor. And Lobachevsky did indeed work at a university outside the center of the world mathematical community. However, this was only true in the trivial geographical

---

3   One neglects contextual factors at one's peril. In particular, it would follow that one would abandon the idea that the inquiry into the origin of knowledge can be reconciled or integrated in a meaningful way. That this is not necessarily so can be adduced from the integration of moderate semantic holism and structural realism (see social integration and holism Chapter 7, based on Esfeld 2001a,b, 2006a,b,c,d, 2007, 2008, 2009, 2010a,b, 2012, forthcoming, Pettit 1993, 2009, Seel 2002).

4   J. Bolyai was the son of one of Gauss's friends, W. Bolyai, who with Gauss attended the University of Göttingen where the parallels postulate was the subject of lectures by Kastner and a number of dissertations. Riemann was Gauss's dissertation student. Lobachevsky worked at a university on the periphery of the European mathematical community, the University of Kazan, which was however staffed by distinguished German professors, including Gauss's teacher, J.M. Bartels. One has to wonder why in the face of the facts of the case Struik and Boyer chose to view things as "remarkable" and "startling." On Gauss's ambivalent role in the development of NEGs, see Robert Osserman's (2005) review of János Bolyai, Non-Euclidean Geometry, and the Nature of Space (MIT Press 2004) by Jeremy J. Gray; published in *Notes of the American Mathematical Society,* 52,9, pp. 1030–34.

sense. The mathematics staff at the University of Kazan included Gauss's teacher, J.M. Bartels and other distinguished German mathematicians.

J. Bolyai's ideas on NEGs were formed as early as 1823. He published his "The Science of Absolute" in a book written by his father in 1832, but the work did not become widely known until the late 1860s (Bolyai's was translated into Italian in 1868). Lobachevsky's works on the foundations of geometry were published beginning in 1825; Riemann's *Habilitationschrift* (supervised by Gauss) was on the foundations of geometry. Gauss corresponded with W. Bolyai (December 17 1799), Taurinus (November 8, 1824), and Besel (January 27, 1829), we know, wrote about NEGs in letters to W. Bolyai (December 17, 1799), Taurinus (November 8, 1824), and to Besel (January 27, 1829) on the topic of NEGs, and published notes on NEGs from 1831 on. The Göttingische *Anziegen* published short reviews in 1816 and 1822. Given the history and density of this NEG network it is curious how distinguished mathematicians and historians of mathematics, including Dirk Struik and Carl Boyer, came to describe the developments as "remarkable" and "startling." Their narratives reflected the widely held assumption that mathematics was a "pure" discipline. Consider, for example, Riemann's case. The traditional history of mathematics narrative is that his generalization of elliptical geometry was an exercise in pure mathematics. There were no practical considerations at work here. But viewed from the perspective a sociological realism,

The idea that there was a concrete possibility of practical applications for this exercise was not a consideration. In the light of a more realistic sociological and network analysis, Riemann's work like that of Gauss, Lobachevsky, Bolyai, Helmholtz, and Clifford was based on the assumption that Euclidean geometry unimpeachable as a game played according to a set of formal rules. However, all of these NEG creators were concerned with the extent to which Euclidean geometry validly represented actual space. In order to determine this the ideas had to be treated scientifically using observation and experiment.

From a sociology of knowledge point of view, the NEG case suggests that behind cases of apparent "individual genius" we should always expect to find a social network. The most impressive argument for this claim is without doubt Randall Collins (1998) comparative historical study in the sociology of philosophies. The rationale here should become clearer over the course of this chapter. In this case, we can also quickly see that the network was grouped around discussions of ideas and problems, for which meetings and letters served as channels (Grosslight, forthcoming) in *constellations* (Muslow/Stamm 2006, Henrich 1952, 1991, 2004, 2008):[5] Kant was interested in the issue of space as a

---

5   Grosslight has redrawn the map around Descartes's situated knowledge in the network of Mersennes. Dieter Henrich's research program on *Konstellationsforschung* unravels the inter-relations of interlocutors sharing a *Denkraum* ("thought-scape") by investigating source materials from reviews, fragments, letters, dialogues that comprise constellations of people and their shared problems of life in light of their philosophical

metaphysical problem, and his correspondence network included the Göttingen circle around Blumenbach and Lichtenberg. At issue was the problematization of extensive and intensive magnitude. Sutherland (2004: 412) supports our claim that one cannot escape the socio-historic context. The distinction between "pure" and "applied" in mathematics as in science generally is a distinction born out of ideological and historico-professional interests (Restivo 1992: 141–6). This is not apparent in the writings of thinkers who conceive of a kind of knowledge that develops independent of space, time, history, and culture. Kant's notion of pure mathematics was more profoundly transcendental. Kant understood pure mathematics in relation to pure space and time. Mathematics is applied to objects of experience in space and time and not to space and time per se. These sorts of arguments can no longer be sustained in the face of the ever increasing efforts to ground ideas in social and material situations and contexts. If we follow through with an idea such as Henrich's *constellation*, we find that it consists of people and their shared problems of life in light of their philosophical ethos, and that the solution of the problems of life and philosophy lay in an ethos that considered itself an "axiomatic method" (Hintikka 2011: 70).[6]

The NEG case is a specific example of the rejection of transcendence, a classic result of Emile Durkheim's study of religion in *The Elementary Forms of Religious Life* published in French in 1912. In this milestone in the history of human knowledge, Durkheim links the social construction of religion and the gods to the social construction of logical concepts. Logic like God seems to have no connection to the everyday/everynight world; such ideas and concepts typically are assigned to a realm of reality outside of time, space, culture, and history. Durkheim's work as a program in the rejection of transcendence (and more broadly of immanence and psychologism) languished until the emergence of the science studies movement in the late 1960 and developments in the sociology of mathematics (notably in the work of David Bloor, Donald MacKenzie, and Sal Restivo).

The "turn to practice" is a notable feature of the postmodern landscape in philosophy and other disciplines. This social turn was anticipated by Courant and Robbins in the their 1906 classic *What is Mathematics* (1995). Mathematics is not a simple matter of consistency and freely conceived ideas but a matter of active experience. Today it is rare but not unusual to find mathematicians who reflect the turn to practice and in some cases the influence of the sociology of mathematics on mathematical thinking (e.g., David and Hersh 1981). In place of classical Platonism, David and Hersh ground mathematics in communal understanding.

---

ethos. See also: Breidbach (2011) on "radical historization" and Stingl (2011) on "semantic agency" and "thought-scapes."

6   *Mannigfaltigkeiten*—which we also find in the Deleuzean concept—is also at the center of the work of Riemann's colleague in Göttingen, Rudolf Hermann Lotze, and his discussion of "the Many and the One," where it attracted the attention of William James and Alfred North Whitehead (in his mereology).

While they still tend to privilege minds and individuals as the creative engines in the creation of mathematics, they exhibit the growing appreciation for the social level of reality emerging across the full spectrum of the disciplines.

Reuben Hersh's riff on the Courant and Robbins classic, *What is Mathematics, Really?* (1997) reinforces the idea of mathematics as active experience and defends an anti-Platonic, anti-foundational perspective. Hersh understands the turn to practice in mathematics as the application of a philosophical humanism rather than a matter of seeing mathematics sociologically. Hersh makes progress toward a sociology of mathematics but his is a weak version. It falls short of the stronger sociology of mathematics as a social construction associated with the science studies movement.

Human beings make mathematics on the earth in societies using the material, social, and symbolic resources available in their cultures and environments. In terms of our goal of reconciling philosophy and sociology in ways that are consistent with the interdisciplinary turn in contemporary research and theory, we welcome and embrace Hersh's philosophical humanism as a step in the right direction.

Mathematics has been a major factor in and product of the unfolding of human civilization. In this section we explore the very idea of civilization as a context for mathematical productions. The ancients contrasted "civilized peoples" with "barbarians;" moderns have opposed "civilized" and "primitive" and "savage." The concept of civilization links mathematics and the idea of progress.

Of course, we utilize the concept *civilization* here in full awareness of its discontents and ambiguities.[7] These concepts, and the concept of *civilization* in particular, reflect a colonial trope and imperialist metaphor that severely limits and constrains the scope of what we have to say—both territorially and temporally—and that recognizes only conceptual realms approved by and subjugated under the Western colonial matrix (Mignolo 2012, also Restivo and Collins 1982; and see Restivo and Loughlin 2000). In full view of the discourse in postcolonial science studies (Harding 1998, 2008, (ed.) 2010), we argue that mathematics is not a monolithic reservoir of absolute and ultimate knowledge. Even from within the Western point of view, it must be acknowledged that mathematics, as a matter of becoming (DeLanda 2002, Stengers and Prigogine 1984, Stengers 1997, 2000), must be reconstructed from a standpoint(s) and situated. Different civilizations (variously "nations," "societies," and "cultures") are associated with different mathematical traditions (v. Restivo 1992: 23–88). To the extent that humans have developed in ways that can be captured in the ideas of "evolution" and "progress,"

---

7   Norbert Elias (1976, 2000) has compared the concepts culture, civilization, civilite, etc. in the new foreword (1976) for his seminal *Civilizing Process* and his later *Studies on the Germans* (and see the classical contributions by: Kroeber/Kluckhohn 1952, MacIver 1917, Rundell/Mennell (eds) 1998; additionally: Anderson 1991, Latour 1993, Linkalter 2007, Merton 1936, Ricoeur 1965).

mathematics, in conjunction with science and technology more generally, is assumed to have contributed positively to and benefited from those developments.

Adam Ferguson (1723–1816) appears to have introduced the term "civilization" as we understand it today in his *Essay on the History of Civil Society* (1767), and perhaps as early as 1759 (v. Benveniste 1954). Boswell (1772), Adam Smith (1776), and John Millar (1771) called on the term in their works. Mirabeau (1757) introduced the term in French in his *L'Ami des hommes ou traité de la population*. Ferguson drew an analogy between the individual development and the evolution of the human species from "rudeness" to "civilization." As the pinnacle of social forms of society, civilizations are characterized by complexity, hierarchies, dynamism, advanced levels of rationality, social progress, and a statist form of government. Not everyone viewed civilization in positive terms. Rousseau, for example, viewed civilization as opposed to human nature.

Ferguson's conception of civilization gives implies that mathematics has been a progressive force in human evolution. Rousseau's conception leads to a much more skeptical view of the virtues of science, mathematics, and civilization.

The idea of progress arguably enters Western culture in the Old Testament's conception of a God that moves through linear time with humans (e.g., Sedlacek 2011: 47). The seventeenth century achievements of Galileo, Newton, Descartes, Molière, and Racine fueled the idea of progress in science, technology, and literature. The Abbé de Saint Pierre (1648–1753) added social progress to the idea of progress in his program for political and ethical academies. His ideas and Turgot's influenced the eighteenth century philosophers who produced the great Encyclopédie under the direction of Denis Diderot. The Encyclopedists defined the Enlightenment's goal of promoting reason and unifying knowledge. In this way, social progress became tied to industrialization and capitalism. The idea of progress in this sense became one with the values and goals of industrial civilization. There was a "dark side" to the Enlightenment that is often ignored when praising it as the very spirit of reason and science. John W. Fleming (2013) has shown that what is sometimes called the "counter-enlightenment movement" is better understood as the co-enlightenment. The esoteric and occult lived on the same paths along which great advances were being made in liberating and liberatory ideas and practices. This is not just another historical case study but a manifestation of the yin and yang of all great movements in the evolution of human reason, liberation, and science.

The standard theories of state formation, power, and the institutionalization of the scientific disciplines traditionally conflate the person-centered account of progress (genius or geometric account (Stingl 2008)) and a Hegelian-Marxist account of power as domination. This conflation has only recently been interrogated from the perspectives of vital materialism and materio-technoscience.[8] There are

---

8   Ahmed 2012, Barad 2003, 2008, Bennett 2010, Chen 2012, Connolly 2011, Coole and Frost (eds) 2010; Dolphijn/van der Tuin 2012, and Mann 1984.

new emphases on military, logistics, and engineering forms of infrastructural power that mobilizes nature in a "figured world."[9]

Veblen, in a locus classicus in the more traditional (geometric) and certainly Western accounts of progress, argued that the various sciences could be distinguished in terms of their proximity to the domain of technology. The physical sciences were closest to, even integral with that domain. Political theory and economics were farther afield.[10] The idea of progress has long been a foil for critics of the way the world works in practice, and now we find ourselves in the middle of an era of unprecedented machine discipline. Unlike machines of the past, the new machines, including social and sociable robots (the so-called robosapiens) and cyborgs, promise to discipline us with conscious awareness and values, thinking and affect. In 1688, Fontenelle published the first modern secular treatise on the idea of progress and the virtues of science as the key to progress. Fontenelle famously introduced the modern notion of time and the idea of the "book of life" as something that *only* needs to be deciphered (Blumenberg 1979, 1986). Rousseau, by contrast, argued that science and the arts have corrupted our minds. They represent the poles of a perennial cultural debate that we have not learned to resolve and that may be an insoluble tension built into the very nature of culture.

There have been attempts to identify a type of progress that is independent of material or technological criteria,[11] but for many ancient as well as modern thinkers the idea of progress has always been problematic. We are right to be concerned about the actual and potential impacts of our new bio- and nano-technologies. But one can find similar concerns in Plato's *Phaedrus*. There, in the dialogue between Theuth and king Thamus concerning the new technology of writing, Theuth makes promising predictions about the impact of writing. The king claims to be in a better position to do what in effect is a "technology assessment," and concludes that writing will have the opposite of the effects predicted by Theuth. The cultural meaning of science has fared no better. Where the Rousseaus and the Roszaks saw danger and alienation in science, the Francis Bacons and Bronowskis saw civilization and progress. In 1923 the biochemist J.B.S. Haldane wrote about a future of human happiness built on the application of science. Bertrand Russell replied with a vision of science used to promote power and privilege rather than to improve the human condition (for a summary and context see: Rubin 2005). St. Augustine worried about the invention of machines of destruction. In the period following World War I, Spengler predicted that humans would be annihilated by

---

9   This is a classic point made by Michel Serres, now in Holland et al. 1998; Bennett/Joyce 2010, Carroll 1996, 2001, 2002, 2012, Joyce 1994, 2003, 2005, Mukerji 1998, 2002, 2003, 2006, 2010, Scott 1998, Stingl 2011b, 2012b,d,e.

10   This is a point that both Patrick Carroll and Alexander Stingl (see also Nico Stehr 1994, 2003) in particular have now taken issue with, following the example of Foucault and Deleuze.

11   For example, see the discussion in Almond, Chodorow, and Pearce 1985, and the classic criticisms in Roszak 1969/1995; or Adorno (1962/64 in:) 2003.

Faustian man. In our own era, and in the wake of Pierre Clastres, Deleuze and Guattari rewrote the State into a war-machine; and C. Wright Mills identified the Science Machine.

Mathematics, like science, is historically enmeshed in the ambivalence about progress. But it stands apart from science in terms of its stronger association with human progress. The historian Florian Cajori (1894: 4) had no question about the connection between mathematics and human progress. For Alex Bellos (2010: ix), mathematics is ("arguably") the foundation of all human progress.

In the seventeenth and eighteenth centuries, a wave of positivism fueled by Newton's achievements evoked the promise of progress among mathematicians of that period. The pedestal the Western academy placed Newton on silenced any earnest talk about Newton the alchemist and theologian. Here again we separate the yin and yang of Newton's life and work at our peril. It requires the courage of a novelist like Neal Stephenson and the scholars who inaugurated "the Newton Project" under the directorship of Sussex professor Rob Lliffe.

Different civilizations (nations, societies, and cultures) are associated with different mathematical traditions (v. Restivo 1992: 23–88). Advocates of the idea of progress assume that all of these traditions have contributed positively to the development of civilization and the progress of humanity.

Progress can mean "amelioration" or "improvement" in a social or ethical sense or it can refer to technological development or evolution. The very idea of progress forces us to be comparative. Are we more advanced than earlier cultures or contemporary ones that are less dominated by machines and machine ideology? How do we measure progress, and how do such measurements whether they are quantitative, qualitative, or intuitive embrace human beings and ecologies? What is the relative value of persons versus property, profits versus sustainable ecologies? Are some cultures more people and ecology friendly than others? The answers to such questions depend in part on the degree to which individuation of the self (and then the myth of individualism, selfishness, and greed) has developed in any given society. In critically evaluating the idea of progress, it becomes quickly apparent that we are obliged to engage values comparatively: people or property, profits or sustainability?

It may be possible to define progress in a way that takes it out of the realm of hopes, wishes, and dreams and plants it more firmly on a meaningful (and even perhaps measureable) foundation. For example, Chester Barnard, in his seminal book on organization, management and leadership titled *The Functions of the Executive* (1938/1968), described humans as fundamentally biological beings constrained mainly by biological and physical boundaries; these limitations could be overcome by cooperation. Gerhard Lenski (1974: 59) defined progress as the process by which human beings raise the upper limit of their capacity for perceiving, conceptualizing, accumulating, processing, mobilizing, distributing, and utilizing information, resources, and energy in the adaptive-evolutionary process. The relationship between adaptation and evolution is a paradoxical one. On the one hand, survival depends on the capacity to adapt to surroundings;

on the other hand, adaptation involves increasing specialization and decreasing evolutionary potential. Adaptation is a dead end. As a given entity adapts to a given set of conditions, it specializes to the point that it begins to lose any capacity for adapting to significant changes in those conditions. The anthropologists Sahlins and Service (1960: 95–7) summarize these ideas as follows:

Principle of Stabilization: specific evolution (the increase in adaptive specialization by a given system) is ultimately self-limiting.

General evolution (progressive advance measured in absolute terms rather than in terms of degrees of adaptation in particular environments) occurs because of the emergence of new, relatively unspecialized forms.

Law of Evolutionary Potential: increasing specialization narrows adaptive potential. The more specialized and adaptive a mechanism or form is at any given point in evolutionary history, the smaller its potential for adapting to new situations and passing on to a new stage of development.

We can add here two additional laws: First, the *Law of Adaptive Levels*: adaptation occurs at different levels across various life orders and systems and occurs at different speeds in different spatial arenas. This law draws attention to the complexity of adaptation and the general processes of variation and selection. And second, the *Law of Agent–Environment Entanglement. Adaptation* suggests an active agent in a stable environment. But active agents can and do change their environments in ways that make different demands on the adapting agents. Looked at another way, environments have agential-like dynamics.

One of the important features of the ideology of science is the makes science (in its mythical pure form) independent of technology. This allows science to deflect criticism onto technology and pin the problem of values on technology. Third world intellectuals seem to have seen through to this aspect of the science and technology nexus more readily than the intellectual Brahmins of the West. As we have turned our attention again to the history of science and technology using the new tools and perspectives of science and technology studies we have revealed intimate connection between what we have traditionally and casually distinguished as science and technology. We have at the same time revealed the intimate connection between technoscience research and development and the production, maintenance, and use of the means (and the most advanced means) of violence in society. It has become clear that the most advanced systems of knowledge are intimately intertwined with the centers of power and production in every society that has reached a level of complexity that gives rise to a system of social stratification.

Perhaps the most important aspect of the ideology of science is that it is (in its mythical pure form) completely independent of technology. This serves among other things to deflect social criticism from science onto technology and to justify the separation of science from concerns about ethics and values. Interestingly, this idea seems to be more readily appreciated by non-western intellectuals than by the

Brahmin scholars of the West and their emulators (cf. Mignolo 2012: Chapter 1; and Prasad 2005, 20006, 2008). Careful study of the history of contemporary Western science has demonstrated the intimate connection between what we often distinguish as science and technology. It has also revealed the intimate connection between technoscience research and development and the production, maintenance, and use of the means (and the most advanced means) of violence in society.[12] Not only that, but this is true in general for the most advanced systems of knowledge in at least every society that has reached a level of complexity that gives rise to a system of social stratification (cf. Žižek 2008).

At the end of the day, it should be clear that progress is not easy to define, and that it is even harder to point to examples of progress that resist critical interrogation. How can we sustain the idea of progress in the face of the widespread ecological, environmental, and human destruction that has characterized the Industrial Age? The fact is that the destruction and danger we see all around us is integrally connected to the very things we use to mark the progress of humanity. For these reasons, we must be cautious when considering whether any of the sciences, engineering disciplines, or mathematics have contributed to or served as signposts of progress. Mathematics, like all systems of knowledge, does not exist in a vacuum. It is always connected to social institutions and under the control of the most powerful institutions in any given society. All of this may put too much of the onus on the sciences and technology when what we are dealing with is culture in general. Is it possible that cultures by their very natures inevitably destroy planets?

Ambivalence about science, technology, and social change (including the idea of progress) and contradictions in the application of science and technology may be built into the very nature of culture. The peoples of the ancient Near East reduced vast forests to open plains through their agricultural activities. Wind erosion and over-grazing then turned those plains into deserts. The Loess Plateau in ancient China was caused by deforestation. The Yellow River, "China's Sorrow," gets its signature color and flooding pattern from the deposits of loess sediment. There is some evidence that the ancient Chinese could have adopted better conservation practices and avoided deforestation and still built China into the greatest civilizational area on earth between the first and sixteenth centuries of the common era. The deforestation experiences ancient world are cautionary tales; we moderns are repeating some of the ecologically unsound practices of the Chinese, the Romans, and other ancient peoples. Even though we are armed with better knowledge and information it is difficult to monitor the multiple emerging and converging complex technologies of our global society. We are living in an era of, technocultural systems out of control.

It is not easy to define progress and even harder to find examples of progress that are not vulnerable to critical interrogation. It is difficult to defend any concept of progress in a world characterized by the widespread ecological, environmental,

---

12   See Langings 2004, Lynn 1997, Mann 2004, 2005, Mukerji 2002.

and human destruction that has characterized the industrial age. Indeed, it is in the human and material destruction we see all around us that we tend to find the very things we use to mark human progress. We would be wise, therefore, to be cautious when we identify the contributions of the sciences, mathematics, and engineering as by definition signposts of human progress. All modes of knowing, all systems of knowledge are connected to and constrained if not controlled by the most powerful social institutions in any given society. We may be putting too much of the burden of progress on science, mathematics, and engineering when the problem is that cultures inevitably destroy planetary systems.

Mathematics and Civilization

The sociology of mathematics can be dated from a number of moments, from Emile Durkheim's remarks on logic as a social fact, or perhaps Marx's remarks on the historical materialism of the calculus, or Oswald Spengler's (1880–1936) views on mathematics and culture. Spengler is a key moment because his position is opposed to that of one of the founders of the sociology of knowledge, Karl Mannheim (1893–1947). Mannheim exempts mathematics from the sociology of knowledge because it is not an ideology, it is not culturally relative, and exists outside of society and history. Mannheim reinforced Pythagorean and Platonist views of mathematics as eternal objects that exist outside of time, space, history, and social life. Until the late 1960s, this view dominated the history, philosophy, and sociology of mathematics.

Spengler argued, by contrast, that each culture is associated with a particular conception of number. His ideas about the "soul" of civilization cannot provide a realistic foundation for a sociology of mathematics. However, his goal of grounding mathematics in particular social and historical forms is nothing less than a sociological and anthropological perspective on numbers. "There is not," Spengler claimed, "and cannot be number as such." There are as many number-worlds as there are cultures; this idea is captured in the statement "There is no Mathematik but only mathematics." Spengler argues that any understanding of the dynamics of civilization requires an understanding of the relationship between number and culture. Number in a "crystallized" culture contributes to that culture's understanding of the human condition and what it means to be human. Mathematics has a special place in culture because it is all at once science, art, and metaphysics.

The terrestrial materialism defended here is opposed to Spengler's spiritualized materialism. Therefore, it is necessary to take some liberties in taking hold of Spengler's ideas in a way that reveals his contributions to the very idea of a sociology of mathematics. Number, like God, represents the ultimate meaning of the natural world. And like myth, number originated in naming, an act that gives humans power over features of their experience and environments. *Nature,* the numerable, is contrasted with *history,* the aggregate of all things that have no relationship to number. It is in this moment that Mannheim and Spengler come close to being at one on the nature of mathematics. Spengler, however, clearly sees the connection between number and culture more clearly than does Mannheim.

Cultures in Spengler's view are incommensurable, and their histories evolve in cycles of change. His general schema of Classical and Western styles and stages in "Culture," "number," and "mind" is essentially an analysis of worldviews. This is reflected in Spengler's attempt to correlate mathematical and other sociocultural "styles." For example, he argues that Gothic cathedrals and Doric temples are *"mathematics in stone."* Spengler is aware of the problem of the limits of a "naturalistic" approach to number and pessimistic about a solution. It is impossible, he writes, to distinguish between cultural features that are independent of time and space, and those that follow from the forms of culture manufactured by humans.

Finally, Spengler claims that a deep religious intuition is behind the greatest creative acts of mathematicians. Number thought is not merely a matter of knowledge and experience, it is a "view of the universe," a worldview. The second claim Spengler makes is that a "high mathematical endowment" may exist without any "mathematical science." For example, the discovery of the boomerang can only be attributed to people having a sense of mathematics that we must recognize as a reflection of the higher geometry.

Sociologists of mathematics have boldly challenged Platonism but have hesitated to follow Spengler. It is difficult even for sociologists to question the apparent self-evidence of number relations. But there are many examples of challenges to the "necessities" of mathematics from inside and outside of mathematics proper. Among the outsiders can be counted Dostoevsky. In his *Notes from Underground* (1864/1918), Dostoevsky argues that $2 + 2 = 4$ is not life but death, impudent, a farce. It's nonetheless "excellent" and we must give it its due, but then we must recognize that $2 + 2 = 5$ is also sometimes "a most charming little thing." Orwell's (1949) also uses these two equalities as metaphoric resources in *1984*. For Orwell, $2 + 2 = 4$ stands for freedom and liberty; $2 + 2 = 5$ stands for Big Brother totalitarianism. Dostoevsky uses $2 + 2 = 4$ to stand for everyday routines and tradition; $2 + 2 = 5$ represents creativity.

Dostoevsky and Orwell are not simply exercising literary privilege. The conventional wisdom on the necessities of mathematics has been challenged by mathematicians and students of the history and philosophy of mathematics. Whether or not $2 + 2 = 4$ is always an empirical question. Where we have long-term experiences with situations in which $2 + 2 = 4$ we are justified in considering those situations closed to further interrogation, that is, we are justified in taking the equality for granted.

The relatively new field of ethnomathematics has provided important data on the culturally situated nature of mathematics and logic. In the 1960s, studies of African mathematics posed intriguing problems for European intellectuals. Robin Horton's (1997) studies from that period on patterns of African thought suggested a different way of reasoning and a different logic by comparison with European thought. Ethnomathematics has helped us to sort out these early discourses and ground differences in cultural patterns rather than in mental proclivities. Malagasy divination rituals, for example, rely on complex algebraic algorithms. Some cultures use calendars far more abstract and elegant than those used in European cultures. Certain concepts about time and equality that Westerners assumed to be universal in fact vary across cultures. The Basque notion of equivalence, for example, is a dynamic and temporal one not adequately captured by the familiar equal sign. Other ideas taken to be the exclusive province of professionally trained Western mathematicians are, in fact, shared by people in many societies.[13]

Ethnomathematics has reinforced the *rationale* for pursuing the Spenglerian program for a sociology of mathematics reflected in authors such as Dostoevsky and Orwell, mathematicians such as Dirk Struik and Chandler Davis, philosophers of mathematics such as Paul Ernest and Leone Burton, and historians of mathematics such as Morris Kline.

Interrogating mathematics in a comparative perspective demonstrates that mathematics develops in different ways at different times and places, identifies the noted mathematicians in given times and places and the social positions they occupied, and reveals their social and communication networks. We also discover the social conditions within and outside of mathematical communities as they go through phases of growth, stagnation, and decline. The degree of "community" among mathematicians, the level of specialization, the extent of institutionalization and the relative autonomy of the social activity of mathematics, it should be stressed, are variable across time and space. Mathematical communities are also epistemic communities (Hass 1989, 1992) that know gatekeepers, civic networks (Baldassari/Diani 2007), discredited communities (Mukerji/Simon 1998), and epistemological vernaculars (Stingl 2011).

It is possible to narrate the history of mathematics in the internalist tradition of the history of science (cf. Shapin 1992) as a more or less linear unfolding that gives the appearance of an inevitable "logical evolution." However, the evidence is that there are a number of variations among the types of mathematics produced in different cultures (cf. Go 2012). The latter "horizontal" variations are *prima facie* evidence for the Spengler thesis. But what about the long-term trends? These too are socially determined, and in two different senses. First, there is a "longitudinal" development in mathematics that unfolds with interruptions, in different cultural contexts, and in stops and starts. Mathematicians move along a certain path at some times and not others. This implies among other things that the concepts of truth and what counts as a proof in mathematics vary over time. A sequence in mathematical development or in the network of mathematicians will stop, start, stop, and start again over a period of a hundred or a thousand years. It will start in one place, stop, and pick up again elsewhere. What drives these processes? And why do particular mathematicians at particular times and places make the major advances and not others? More to the point, how is that

---

13   See Ascher 2004, D'Ambrosio 2006, Eglash 1999, Mesquita, Restivo, and D'Ambrosio 2011, Verran 1992.

some mathematicians and not others who might be better candidates get credit for discoveries and inventions?

Various social factors affect the variations, interruptions, progressions, and retrogressions characteristic of the longitudinal development of mathematics. Much of professionalized mathematics is created within the boundaries of the mathematical community. Internal competitions and challenges fuel the development of higher and higher levels of self-awareness among mathematicians about their practices. The result is what we conventionally view as increasing levels of "abstraction." Sociologically, this is the consequence of higher levels of specialization and institutional autonomy among mathematicians. The Spengler thesis, then, is true in a very strong form: "number," and all that it stands for metaphorically, is a *socially created activity,* or more technically, a social construction.

We should consider replacing the term "abstraction" with the term "generalization" if for no other reason than that "abstraction" has very fertile and historically concrete meanings in the works of Kant, Whitehead, and Deleuze. We offer alternatively a rationale grounded in sociological materialism. Abstractions are simply concrete forms constructed under conditions of professionalization and disciplinary closure. As mathematics becomes more organized and disciplined, mathematicians build new levels of mathematics on the grounds of earlier mathematical forms. This removes mathematics further and further *ceteris paribus* from the everyday world and mathematics takes on the appearance of something increasingly abstract. It is important to see these developments in material terms. "Abstraction" makes us vulnerable to the myths and ideologies of purity and even to ideas about heavens and gods (cf. Stingl and Weiss 2013).

Selected Case Studies

The world history of mathematics has not unfolded in a unilinear, unidirectional manner. The Greeks, for example, took a step backward from the Babylonian achievements in notation Different types of mathematical systems have developed in different parts of the world; and rival forms of mathematics have sometimes developed *within* societies and professional networks. Hindu mathematics prior to the introduction of Greek astronomy around 400CE was uniquely based on large numbers. This mathematics emphasized the use of numbers in sociocultural schemas and ignored classical practices in geometry, arithmetic, number theory, and algebra. The *Upanishads* (ca. 700 to 500BCE) are full of numerological descriptions: 72,000 arteries; 36,360, or 36,000 syllables; the 33, 303, or 3306 gods; the 5, 6, or 12 basic elements out of which the world is composed. Buddha wisdom is illustrated by the gigantic numbers he can count out (on the order of 8 times 23 series of 107*),* and his magnificence is shown by the huge number of Bodhisattvas and other celestial beings who gather to set the scenes for his various sutras. Hindu cosmology includes cyclical views of time enumerating great blocks of years called *yugas*. There are four *yugas* ranging from 432,000 to 1,728,000 years, all of which together make up one thousandth of a *kalpa* or 4,320,000,000 years.

Hindu cosmology gives us a distinctive view of the infinities of being that exist beyond our immediate terrestrial experiences. It should not be surprising then the Hindus should have invented the zero around 100CE (*sunya* in Sanskrit, meaning literally "emptiness"). *Sunya* was the central concept in Madhyamika Buddhist mysticism, and preceded the invention of the mathematical zero about 600CE. Brahmagupta published a number of rules governing the use of zero and negative numbers in his *Brahmasputha Siddhanta* (ca. 630CE). Classical Indian worldviews are permeated with a "mathematics of transcendence" by contrast with mathematical work concerned with rationalistic generalization. Numbers were used in numerological rhetorics to mystify, impress, and awe. The social roots of this distinctive system lie in the exceptionally exalted status of Indian religious specialists. The roots of the Hindu emphasis on large numbers lies in the great variety of ethnic groups making up Indian society and institutionalized in the ramifications of the caste system.

Chinese mathematics is expressed in concrete pictures. The cosmological significance of Chinese mathematics is rooted in this ideographic bias. The *I Ching*, the ancient book of divination based on a system of hexagrams, was continuously reinterpreted in successive Chinese cosmologies to reflect the changing universe. Chinese arithmetic and algebra were worked out in positional notation. Different algebraic unknowns were represented by counting sticks laid out in different directions from a central point. Chinese algebra, at its height around 1300CE, could represent complex equations and included the idea of determinants (i.e., the pattern of coefficients). The ideographs (and the social conditions of their use) helped preserve the everyday roots of mathematics and obstructed the invention of increasingly general forms of mathematics. This underscores the significance of the professionalization of mathematics in the West for developing increasing levels of generalization.

The inertia of Chinese mathematics was integral with the maintenance of ideographic writing among Chinese intellectuals. Together they gave Chinese culture a concrete aesthetic form. The ideographic form had technical limitations that a more generalized form—an alphabet, a more mechanical mathematical symbolism—would have overcome. Ideographs are hard to learn; they require a great deal of memorization. These limitations may in fact have been the reason why Chinese intellectuals preferred them. A difficult notation is a social advantage to a group attempting to monopolize intellectual positions. This may be contrasted with the algorithmic imperative characteristic of periods of rapid commercial expansion.

Individuals and groups who possessed the esoteric skills of writing and mathematics in the ancient world tended to be state or religious dignitaries. Learning the notational systems of writing and mathematics required long and difficult periods of study by elites and thus were highly resistant to change. Sanskrit, for example, was written without vowels and without spaces between the words. Egyptian writing was similarly conservative. Chinese writing and mathematics are notable because archaic styles lasted much longer than anywhere else. The over-riding cultural issue might be that China was, as Leon Stover argued (1974: 24–5), a "once and always Bronze Age culture," the only primary civilization to develop its Bronze Age to the fullest.

The development of ideographs and mathematical notation in China was in the direction of greater aggregative complexity and aesthetic elaboration, not of simplification and generalization. The Chinese literati thus managed to make their tools progressively more difficult to acquire. This is in keeping with the unusually high social position of Chinese intellectuals. They maintained their status through an examination system that was used to select officials in many dynasties. Many students of mathematics have contended that a "good" notation is a condition for progress in mathematics. For any given period in the history of mathematics, we are faced with the question of why a more appropriate symbolism wasn't invented. The answer is rooted in struggles between monopolizing and democratizing forces over access to writing and mathematics.

Monopolistic groups tend to be strong in highly centralized administrations such as ancient Egypt, the Mesopotamian states, and China. Decentralized administrations foster democratizing forces and independent commercial activities; this was the case, for example in ancient (especially Ionian) Greece, and certain periods in ancient and medieval India. Countervailing forces are a feature of the more developed states. Specific social institutions and practices in developed states will also exhibit countervailing forces. Greek mathematics had some conservative elements, especially in the Alexandrian period when difficult rhetorical forms of exposition limited the development of algebra. The specific character of mathematics in given world cultures is due to the differential incidence of such conditions.

Greek mathematics, the font of modern mathematics in the West emphasized geometry, generalized puzzles, and formal logical proofs. During the Alexandrian period, a numerological arithmetic was developed and applied to revealing a mystical cosmology. Numbers were associated with Hebrew or Greek letters, allowing any word to be transformed into a number that in turn would reveal mathematical relations to other words.

Numerology is related to Hebrew Cabbalism, Christian Gnosticism, and the Neo-Pythagorean revival associated with Philo of Alexandria (ca. 20BCE–50CE). The most prominent expositor of this new mathematics was Nichomachus (ca.100CE). Like Philo, he was a Hellenistic Jew (living in Syria). It was in this Jewish-Greek intellectual milieu of the Levant that the major religious movements of the time emerged.

These types of variations continue to show up in modern European mathematics. There are conflicts between alternative notational systems in the 1500s and 1600s; and a century-long battle between the followers of Newton and those of Leibniz over the calculus. In the nineteenth century, a major dispute arose between

Riemann, Dedekind, Cantor, Klein, and Hilbert and critics such as Kronecker and Brouwer. This split continued and widened in the twentieth century. The result was the emergence of schools of formalists, intuitionists, and others in conflict over the foundations of mathematics.

The Social Roots of Mathematics

The basic activities of everyday survival in the ancient world led to the development of the two major modes of mathematical practice, arithmetic and geometry. The development of arithmetic is stimulated by problems in accounting, taxation, stock-piling, and commerce; and by religious, magical, and artistic concerns in astronomy, in the construction of altars and temples, in the design of musical instruments, and in divination. Geometry is the product of problems that arise in measurement, land surveying, construction and engineering in general. Arithmetic and geometrical systems appear in conjunction with the emergence of literacy in all the earliest civilizations—China, India, Mesopotamia, Egypt, and Greece. These mathematical systems are, to varying degrees in the different civilizations, products of independent invention and diffusion. While we can generally analytically identify and distinguish what is arithmetic and what is geometry, this is not always easy to do in practice.

Arithmetic, geometry, number work, and some form of proto-mathematics are found in cultures throughout recorded history. The discipline of mathematics emerges and becomes institutionalized under specifiable social and ecological conditions. The general social organizational and human ecological conditions for the emergence of modern science are discussed in Restivo and Karp (1974; and see Restivo 1979, and 1994: 29–48). As those conditions crystallized in Western Europe beginning in the 1500s and earlier, the discipline of mathematics emerged as sets of arithmetic and geometrical problems were assembled for purposes of codification and teaching, and to facilitate mathematical studies. Assembling problems was an important step toward unifying mathematics and stimulating generalization. Institutionalization is the precondition for generational continuity, and generational continuity is the necessary condition for the sustained productions of a discipline. This is illustrated from an epistemological point of view in Foucault's *The Order of Things* (2002), for biology in John Dupre's *The Disorder of Things* (1997), and from a material practical perspective in Chandra Mukerji's "Unspoken Assumptions" (1998).

Several things were necessary to unify, institutionalize, and discipline mathematics. One was the effort to state general rules for solving all problems of a given type. Once problems were arranged so that they could be treated in more general terms, they could be detached from their practical origins and addressed as hypothetical puzzles. At this stage problems could be invented without explicit reference to practical issues. The three famous puzzles proposed by Greek geometers of the fourth and fifth centuries BCE are among the earliest examples of such puzzles: to double the volume of a cube (duplication of the cube), to construct a square with the same area as a given circle (quadrature of the circle), and to divide a given angle into three equal parts (trisection of the angle). Such problems were related to the non-mathematical riddles religious oracles commonly posed for one another. One account of the origin of the problem of duplicating the cube, for example, is that the oracle at Delos, in reply to an appeal from the Athenians concerning the plague of 430BCE, recommended doubling the size of the altar of Apollo. The altar was a cube. The early Hindu literature already refers to problems about the size and shape of altars, and these may have been transmitted to Greece by the Pythagoreans, a secret religio-political society. The problem is also a translation into spatial geometric algebra of the Babylonian cubic equation $X^3 = V$.

The duplication, quadrature, and trisection problems fueled the debate oriented Sophists. Plato later introduced the constraint that the only an unmarked straightedge and compass could be used in solving these problems eliminating specialized mechanical devices from us in mathematical competitions. This stiffened competitive conditions and put an emphasis on intellectual means and "gentlemanly" norms. Plato's Academy was organized to support the political ambitions of elite intellectuals who supported an aristocratic state opposed to democracy and commerce. These intellectuals developed an ideology of extreme intellectual purity, glorifying the extreme separation of hand and brain classical Greece's slave economy.

The three Greek puzzles and other problems facilitated competitive mathematical games of challenge-and-response. Such competitions became important moments in the development of Western mathematics. Up until the nineteenth and twentieth centuries, patrons, academies, and states initiated, endorsed, or rewarded competitions, often offering prizes for solutions to practical problems. These competitions reflected economic problems, the prestige of governments along with individual and group struggles for intellectual preeminence.

The initiation of mathematical contests in ancient Greece was accompanied by the rise of a formal, logical mode of argumentation that built on earlier methods of proof. This led to systems of interrelated proofs culminating in Euclid's *Elements* around 300BCE. Euclid published a collection of problems along with definitions, postulates, and what we recognize today as axioms. Euclid, like Aristotle, did not use the term "axiom" but something closer to "common notion." Their work generalized and codified vast eras of human experiences. The Aristotelian and Euclidean innovations in systematization-and-generalization introduced one of the two major generators of new mathematics. The other major generator was applied mathematics.

The empirical path to new mathematics involves applying existing mathematics to new and old practical problems. Most of the early Greek geometrical puzzles, for example, concerned flat figures. Plane geometry could be easily extended to solid spherical geometry as well as to the geometry of conic sections. The geometry of conic sections eventually led to work on curves of various shapes. These developments were carried out during intermittent periods of creativity in Alexandrian mathematics (especially from 300 to 200BCE and 150–200CE). A number of new specialties appeared during this era, but no new level of generalization appeared with the important exception of trigonometry.

There are many parallels between the histories of arithmetic and geometry. Generalizations that arose from solving numerical problems led gradually to algebra. Challenge and-response played an important role in these developments. For example, there is this famous problem, attributed to Archimedes (287 to 212BCE): find the number of bulls and cows of various colors in a herd, if the number of white cows is one third plus one quarter of the total number of black cattle; the number of black bulls is one quarter plus one fifth the number of the spotted bulls in excess of the number of brown bulls, etc. Problems like this with their unknown quantities, led over time to various kinds of notations and symbolisms. These took quite different directions in ancient and medieval China and India, the Arab world and later in medieval and Renaissance Europe. The creation of a highly generalized symbolism that could be easily and mechanically manipulated in problem solving began to crystallize in the late 1500s and 1600s in Europe.

Various forms of algebra evolved across civilizations by way of empirical extension over many centuries. One method that aided in this evolution was the deliberate invention of problems with ever increasing unknowns of higher and higher powers. Equations of the form $ax + b = c$ gave way to those on the order of $ax^4 + by^3 + cz^2 = g$. There was in principle no end to the complexity of these equations. Vieta in the 1580s, for example, was challenged to solve an equation involving x^{45}; but the extensions also gave rise to efforts to find general rules for solving higher order equations. Such empirical extensions promoted the search for generalized extensions. At the same time, arithmetic was developing in other directions.

Elementary arithmetic was characterized during these periods by a tremendous variety symbols and rules, more and less difficult in terms of their use in solving practical problems especially when working with large numbers, fractions, or complex operations like division or the extraction of roots. Problems were usually expressed in words. A great deal of creative effort went into developing readily manipulated notations. The invention of decimal place notation and the zero sign in India were among the most important results of these efforts. Other important innovations were the standardization of positional methods in multiplication and division (in Europe ca. 1600); and the invention of logarithms by the Scotsman Napier in 1614, for use in astronomy, navigation, and commerce.

In another line of development in arithmetic, the properties of numbers themselves became a focus of interest for Eratosthenes (ca. 230BCE). He was an earlier leader in the effort to develop a general method for identifying prime numbers. The Pythagorean work on "triangular' and "square" numbers anticipated Fermat's theorem that every prime number of the form $4n + 1$ is a sum of two squares. These early efforts in what we now know as number theory occupied occultists and cabalists during the Alexandrian period. The more conventional puzzle solving forms of number theory continued to arrest the attention of number workers and mathematicians up to the Renaissance and from then into the contemporary period.

During the Alexandrian period, Hipparchus (ca. 140BCE), Menelaus (ca. 100BCE) and others developed a new branch of mathematics based on a combination of arithmetic and geometry. Trigonometry evolved from the measurement of angles and lines and ratio calculations and spread to medieval India and the Arab world. When it reached Renaissance Europe it became the foundation on which Napier invented logarithms.

What we find, then, is that mathematics arises from everyday activities involving measurement and calculation. The systematization of these activities leads to geometry and arithmetic. Competition and puzzles helped to generalize number work and the development of mathematics as a discipline. Geometry and arithmetic become recognizable forms of number work and mathematics and give rise to Euclidean and eventually non-Euclidean geometries, and early on to trigonometry. The generalization of arithmetic leads to more and more complex algebras and number theory. New fields continue to emerge as Europe moves from the middle ages through the Renaissance and continues into our own times. These developments were the result of generalizing generalizations (commonly understood in terms of increasing levels of abstraction), applying generalizations to new practical problems and puzzles, and creating hybrid fields. Descartes and Fermat led the way in combining algebra and the new coordinate geometry to produce analytic geometry.

Problems of motion and curvilinear dynamics led the early Greeks to initiate analyses that would eventually lead to the development of the calculus. The Greeks invented the method of exhaustion in the fifth century BCE for determining the area under any arbitrary shape. Antiphon is credited with originating the idea without fully grasping it. Some decades later, Eudoxus of Cnidus put the idea on a rigorous footing. The method involves inscribing a series of polygons inside the given area. The areas of the polygons converge to the area of the shape. Correctly constructed, the difference in area between the n-th polygon and the containing shape will become arbitrarily small as n becomes large and we approach the actual area of the shape. The term "method of exhaustion" wasn't introduced until the middle of the seventeenth century in a work on geometry by Grégoire de Saint-Vincent (*Opus geometricum quadrature circuli et sectionum* 1647). The method of exhaustion, which requires applying the reductio ad absurdum, is a precursor to the development of the concept of limit in the calculus that evolved in Europe from Newton and Leibniz onward into the nineteenth century and even into our own era

Calculus was applied to successively more complex functions (empirical extension); and eventually (in the 1800s) it was generalized into a theory concerning such things as the rules for solving equations, and the properties of all functions (generalized extension). The drive towards creating new fields by generalization and extension is something we notice in highly competitive periods. Geometry itself experienced a rapid series of branching around 1800 and thereafter, the best known being the non-Euclidean geometries. Other developments included descriptive geometry (Monge) projective geometry (Poncelet), higher analytical geometry (Plucker), modern synthetic geometry (Steiner and Von Staudt), and topology (Möbius, Klein, and Poincaré). Klein, Hilbert, and Cartan unified these different geometries in the late 1800s and early 1900s. The most important moment in this unification was Klein's Erlangen Program. Cartan generalized Klein embedding the unification into and its generalization in Cartan's program, which was designed to place the unification into the framework of Riemann's geometry.

There were parallel developments in algebra after 1800.The effort to find a general solution for the quintic and other higher-order equations led to the creation of the theory of groups by Abel, Galois, Cauchy, and others. This theory focused on an abstract pattern among the coefficients of equations, and opened up a new area of inquiry in higher mathematics. "Abstract" algebras were created by Boole, Cayley, Sylvester, Hamilton, and Grassman. All of these new tools were applied to other branches of mathematics. Dedekind applied set theory to the calculus, Cantor applied it to the concept of infinity, and others applied it to topology, number theory, and geometry. These developments led to the creation of yet another even more general field toward the end of the nineteenth century, "foundations." "Foundations" focused on the nature of mathematical objects themselves and with the rules by which mathematics should be carried out. Foundations research has been the focus of a number of opposing schools, and has led to what are probably the most intense controversies in the history of mathematics.

To review the basic narrative once more, arithmetic, geometry, and other mathematical forms arise from practical problems in measurement, weighing, construction, taxation, administration, astronomy, and commerce. The practical stimulus to number work and mathematics persists throughout the history of mathematics. In the pre-modern period, ballistics and navigational problems promoted the development of the calculus. In the context of the industrial revolution, the requirements of new technologies were answered by descriptive geometry and Fourier's analysis. The conclusion we can draw from a review of this history is this: an increase in the amount, type, intensity, or scope of practical concerns in a society will stimulate mathematical activity. Commerce and activity in the economic sector tends to be very stimulating for mathematics. Mathematical innovations are also stimulated by the introduction of new technologies of production, warfare, transportation, and new modes of administration. Modern European mathematics is intricately intertwined with the emergence and unfolding of modern industrial technological societies. These societies are often labeled "capitalist." It is important to point out that the term "capitalism" does not refer to an actual economic system but rather expresses an economic ideology. Since this is one factor among several, it does not imply that mathematics must come to an end in non-capitalist societies. It does, however, suggest that the form and content of mathematics (within the constraints noted by Spengler) as we know it today is a product of specific lines of cultural development. One could say that modern mathematics, like modern science, is part of the knowledge system generated by and supportive of capitalism.

Mathematics and Social Formations

While a variety of mathematical traditions addressing a variety of practical concerns are features of the comparative history of mathematics. However, certain mathematical forms tend to dominate in given cultures or in certain cultural periods. The history of Chinese mathematics is dominated by an inductive "mathematics of survival." The origin narrative of Chinese mathematics begins with the myth of Yü the Great Engineer's discovery of a magic square on the back of a Lo River tortoise (ca. 2200–2100BCE). From this mythical origin, Chinese mathematics reached its "Golden Age" in the late Sung and early Yuan dynasties. One of the classics in the history of mathematics comes from this period, Chu Shih-Chieh's "Precious Mirror of the Four Elements," written in 1313.

The history of Chinese mathematics is a history of dealing with everyday problems such as taxation, barter, canal and dike construction, surveying, warfare, and property matters. Chinese mathematics didn't reach a generalizing stage because the Chinese social formation couldn't support the organization of an autonomous mathematical community with generational continuity.

Ancient Greece supported a relatively more autonomous philosophical community that embraced mathematics. Conditions in ancient Greece were more favorable for generalizing mathematics. The commercial expansion in Greece in the 600s BCE stimulated mathematical growth. Learned merchants practiced and taught mathematical arts, and master-student relationships across generations (generational continuity) fostered mathematical progress. In consort with these developments, Greek political economy was moving in the direction of an increasingly elitist, self-aware, and self-perpetuating intellectual community. The eventual result was the oligarchic society of Plato's era. The "thinking Greeks" thrived under conditions that divorced hand and brain. A leisure thinking class evolved out of but within institution of the philosopher-merchant number workers. What we know as classical Greek mathematics came out of the slave-based class structure of the post-Ionian period. Arithmetic developed among the slaves who were responsible for most commercial activities, and among householders carrying out the calculation of everyday living. Geometry became the province of the elite thinking class. Geometry was considered democratic but also more relevant to ruling class interests.

Let us now consider some general principles that arise out of the study of the division of labor and specialization. If specialization within the division of labor develops unchecked, it fosters virtuosity. As a result the virtuosi become removed from the problems and social practices of everyday life. There work tends increasingly to be generated by problems from within the social boundaries of the specialization. Those problems are increasingly formulated in terms of the specialized language (symbols) of an increasingly professionalized community. The result is that solutions become networked increasing levels of generalization (results become in more conventional terms more abstract). The other consequence of these developments is that they generate ideologies of purity. This was what happened within certain limits in classical Greece, limits imposed by the limits of political, economic, and cultural growth. In the wake of Plato's death we already witness the reuniting of hand and brain. Aristotle shows a clear interest in linking mathematics and everyday problems. Hand and brain were basically united during the Alexandrian period but the ideology of purity remained viable. Archimedes work, for one notable example, reflected the unity of hand and brain but he championed a purist philosophy. The decline and fall of Greek commercial culture brought an end to Greece's vital Greek mathematical culture. Archimedes brought Greece to the threshold of the calculus and his work marked the zenith of Greek mathematics. Mathematics was revived during the European commercial revolution beginning as early as the twelfth century. The coming economic revolutions that we know of as capitalism were signaled by the building of the Gothic cathedrals. These edifices were made possible by the accumulation of the technologies of labor and construction. As capitalism emerged, problems of motion and change challenged merchants, navigators, and merchants of war and called forth further developments in the calculus rooted in the contributions of Eudoxus and Archimedes. Their works were part of the recovery and spread of the ancient Greek texts notably by way of the Arabs. From the late 1600s on Newton and Leibniz reduced the basic problems of rates of change, tangents, maxima and minima, and summations to differentiation and anti-differentiation (indefinite integration). Studies of motion led to the development of the idea of a function. The idea of a function played a central role in the innovations of seventeenth- and eighteenth-century mathematics. The intimate relationship between theology and mathematics is exemplified in the history of the calculus. Infinities and infinitesimals were central topics of debates among theologians and scholastics in the centuries leading up to Newton and Leibniz when they into the construction of the calculus and the development of capitalism. The Euclidean realm of the straight, flat, and uniform gave ground as a new dynamic world of skews, curves, and accelerations marking the increasing prominence of technologies of war and commerce. The search for algorithms, time-saving rules for solving problems, is evident in the writings of the inventors of the calculus (e.g. in Leibniz's "De geometria recondita et analysi indivisibilium atque infinitorum" of 1686).

The machine of the calculus co-evolved with the machine of capitalism as did the machine of analytic geometry that Descartes played a central role in inventing. The calculus and analytic geometry (described as an industrial process by Boutroux 1919) were both driven by an algorithmic imperative. The two mathematical systems were in constant association notwithstanding the conflict between Cartesians and Newtonians. Incidentally, Newton and Leibniz invented two different calculuses. Newton's "method of fluxions" was heavily indebted to classical Archimedean mathematics. One of the main advantages of Leibniz's "differential calculus was its notation (Restivo 1992: 134).

The machines of the calculus and analytic geometry transformed mathematics into manufacturing. Side by side with the Greek commercial revolution and the emergence of capitalism in Europe we can consider what happened as a result of Japan's seventeenth century commercial revolution. The Japanese developed a native calculus as their commercial revolution unfolded and a monetary economy was established. These developments were fueled by contacts with Europeans at Japanese ports of entry.

Puzzles and Proofs

From the earliest times in the development of mathematical practices and institutions, puzzles and competitions have helped to drive mathematical growth and development. This has been an especially prominent feature of mathematics in the West. There have been periods in the history of Western mathematics dominated by public challenges. During the thirteenth century, Leonardo Fibonacci (ca. 1200) was issued challenges by Emperor Frederick court mathematician; Tartaglia and Cardano challenged one another in sixteenth century Italy; public challenges led to high acclaim for Vieta at the French court in the 1570s. Puzzle contests and public challenges played an important role in driving mathematics to become more general. The difficulty of such contests and challenges was ratcheted up by problems that had nothing to do with the problems of everyday life. Notable examples include the contests that led Tartaglia to solve cubic equations and motivated Vieta to figure out how to reduce the search for general solutions to equations, such as those that Tartaglia found for cubic equations and Vieta found for the reduction of equations from one form to another.

An emphasis on *proofs* has characterized various periods in the history of mathematics, provoked by competitive engagements. The Greeks rationalized the construction of proofs during the period that mathematics became popular among a class of elite philosophers and competition for power, status, and position was high in the intellectual world. It was during this time that wandering Sophists were engaging in debating contests that were arenas from which canons of logic began to emerge. There were parallel developments in terms of causes and effects in debating, logic thinking, and mathematics embodied in the Sophists who counted among themselves many of the mathematicians of the time. Many of the formal schools organized in this classical period, such as the Academy, relied on mathematical skills as proof of their superiority over competing institutions. A stress on proofs could clarify the rules of competition and escalate their intensity. Most of the inventiveness of Western mathematics is rooted in mathematical competitions and the challenges posed by ever-new puzzles.

The preceding internalist analysis must be understood as taking place with a larger context of economic stimuli. The case of Thales is iconic. In the first place he might be a composite or imaginary character. Real or imagined, he symbolizes the role of the philosopher-merchants who developed more general forms of proof than those invented by the Egyptians and Babylonians. Thales reflects the need among the commercial Ionians to understand the physical and natural world and to have at hand highly developed computational methods. Commercial interests and goals stimulated the development of proofs that were crude extensions of Babylonian or Egyptian "rules" for checking results. As proofs became increasingly systematized and rationalized they evolved over the next 300 years into the forms of proofs we find in Euclid's geometry.

In China and India, mathematics was less organized than in the post-Renaissance West, more episodic (especially in the case of India), and more focused on a "mathematics of survival," proofs were almost entirely ignored. In some cases, problems would be stated without solutions or with solutions that were incorrect. The lack of competitions and challenges in number work and mathematics goes a long way toward explaining why proofs were not important in these civilizations. The other important factors in accounting for the limited development of proofs are (1) the low density of the networks of mathematical and number workers, and (2) the fact that most mathematical and number workers in the East were governmental officials or otherwise functionaries in state bureaucracies. By contrast with the situation in classical Greece and modern Europe, there are never more than a few mathematicians or number works working in any given period. The social networks of number workers and mathematicians in creative periods especially were very dense in classical Greece and modern Europe. These networks were made up of private individuals and/or teachers in competitive itinerant or formal school systems. Their counterparts in the East were functionaries and insulated from competitions. Most of the Oriental mathematicians were government officials and thus were insulated from outside competition.

Given the preceding discussion we should not be surprised to find a limited emphasis on proofs in the Islamic-Arabic world, which experienced a flurry of mathematical activity in the period 800–1000 (and later to some extent). There was some concern for proofs (in the works of Thabit Ibn Qurra and other scholars showed some interest in proof but nothing like that we find in classical Greece. The level of interest they showed in proofs was to some extent the result of translating the Greek works that had come into their hands. But they lacked a densely populated community of number works and mathematicians, competition was relatively underdeveloped as were master-student chains, and generational continuity in the Golden Age was limited.

Thus, proof work, competitions, and generational continuity were limited in India by its episodic history of mathematics, in China by its Bronze Age dynastic structure, in Islam by the boundaries and structure of its Golden Age. Finally, Japan's seventeenth-century mathematical and commercial renaissance are ended when the Tokugawa regime comes to power (Restivo 1992: 22–60). Only in Europe from the 1500s on do we find a sustainable generational continuity and population density capable of supporting long-term mathematical growth and development.

Modern European mathematics has witnessed a steady growth in the emphasis on and nature of proofs. The seventeenth century mathematician Fermat did not accompany his theorems with proofs. Every student of the history of mathematics is familiar with the marginal note in Fermat's copy of Diophantus' *Arithmetica*: in respect to "Fermat's Last Theorem" that there is no solution for the equation $a^n + b^n = c^n$ for any integer value of n greater than two, he writes that he has "a truly marvelous demonstration of this proposition which this margin is too narrow to contain." In the eighteenth century, Euler, one of the most prolific mathematicians in history constructed proofs that were characteristically not very rigorous. More rigorous standards for proofs were introduced beginning the early 1800s saw and this led to a rejection of many earlier solutions. These solutions may have been formally correct but in the new context were judged to be based on insufficiently comprehensive and universal reasoning. The increasing rigor of proofs and of mathematics in general came along as the number of mathematicians grew significantly as educational systems, notably in France and Germany, expanded. Improved standards of proof and a shift to more rigorous requirements pressed mathematics to new levels of generalization. Proofs need more general elements than specific numerical exemplars, and new standards of rigor prompted increasing attention to the very nature of—the foundations of—mathematics.

Reflection and Generalization inside Mathematics

Psychologically, one can narrate the history of mathematics as one of increasing reflection and self-awareness among mathematicians. At some level, mathematicians lost their Platonic innocence and their more terrestrial naïve realism about mathematician objects. A key moment in this process occurred when mathematicians stopped dropping negative roots of equations (standard practice among Hindu, Arab, and medieval European mathematicians) and started to use them. In spite of the Hindu antipathy to negative numbers, Brahmagupta stated some rules for dealing with them in the seventh century. But as late as 1758 the British mathematician Maseres described negative numbers as darkening the field and complicating issues that were simple and obvious. Beginning in the early nineteenth century, DeMorgan and others helped incorporate negative numbers into the everyday practices of mathematics as they constructed and systematized the logic of arithmetic. Later they recognized that imaginary numbers (an unfortunate and distracting nomenclature) were useful despite their apparent absurdity. Gauss established a new basis for modern algebra by creating a representational system for complex numbers. Nineteenth century higher mathematics took off from this point.

The key reflection was that mathematical concepts could be deliberately created. Mathematics was not completely constrained by the material world common sense representations of that world. The self-conscious creations of mathematics were never *ab novo* but drew on cultural objects and ideas in the wider society and within the professional communities of mathematics. This new self-consciousness was popularized by the creation of new generalized geometries (including projective and non-Euclidean geometries). New algebras and more general forms of analysis soon followed.

Mathematical objects are real in the sense of operations as opposed to things. The imaginary number *i* symbolizes an activity, the *operation* of extracting a square root from a negative number. The fact that this operation cannot actually be carried out is no obstacle to it mathematical utility. Mathematicians are used to working backwards from unknown solutions to premises by symbolizing the solution using an arbitrary designation (e.g., *x*). *i*, then, can be used as the basis for mathematical operations that can actually be carried out. Operations come in different degrees of complexity; some examples are the basic arithmetic operations, the concept of a function, and the concept of a group. Every natural whole number is an operation—the operation of counting. Other operations may be involved and this is one of the questions modern mathematicians are looking into.

Modern mathematics develops by taking new operations as new units of analysis. These new units can be crystallized in new symbols that can then be manipulated as things, matters of fact, the material resources of the mathematical community (see Aleksandrov, Kolmogorov, and Lavrent'ev 1963 for a thoroughly materialistic and realistic history of mathematics).

Mathematics grows by treating operations as objects that can in their turn become the subject of new operations. The self-awareness about generalizing operations contributes to a novel form of reification and a process commonly understood as moving up levels of abstraction. The symbolic focus of Western mathematics is a function of a social network of mathematicians achieving higher and higher degrees of professional self-awareness and reflexivity. This development crystallized in the late nineteenth and early twentieth century and led to a concern for foundations but also to an increasing realization that mathematics is a contextual, situated human, and in particular *social and cultural*, practice. The fact that there has been a turn to practice and experience in mathematics and the philosophy of mathematics has not been enough to eliminate Platonic thinking among mathematicians and logicians.

Specialization in mathematics, like specialization in other modern activities, has reached levels unknown in earlier historical periods. The more specialized mathematics has become the more mathematics itself has become the generative force behind new mathematical ideas. This has meant that the generative force of activities and ideas outside of mathematics proper has diminished. The institutional context of professional, bureaucratized mathematics has the effect of isolating mathematics from subsistence production in the society at large.

We now have a sociological platform on which to consider a classical philosophical and historical problem, captured in the title of a famous paper by

Eugene Wigner (1960) on "The Unreasonable Effectiveness of Mathematics in the Natural Sciences."

The fact that "pure" mathematics turns out to be useful after the fact in the physical and natural sciences is not a mystery at all. The mystery is the fallout of two fallacies. The first fallacy is the fallacy of purity. Mathematics, as we have seen, does not exist in a "place" outside of time, space, history, and culture. The second fallacy is the fallacy of unreasonable utility. The fact that mathematics for its own sake (mathematics generated within and for the mathematics community) turns out to be useful in the sciences is neither mysterious nor a coincidence. It reflects wider and deeper truths about mathematics and science. The utility of mathematics for science arises from the simple fact of a constant interplay between mathematics and science. Those who find the effectiveness of mathematics unreasonable should pay more attention to the interruption of the interplay between mathematics and science. This interplay can be and has been interrupted by the professional and bureaucratic closure of the mathematical community as a social system. The greater the degree of closure—that is, the more advanced the processes of professionalization and bureaucratization get—the less likely we are to observe the effectiveness of mathematics for the sciences. Closure will interrupt the cycles of effective applications outside of mathematics per se. This is not just a sociological conclusion based on the theory of closure and autonomy in social systems (Restivo 1992: 171–5, Restivo 1993: 263–7) but something that has been recognized by professional mathematicians (e.g., Boos and Niss (eds) 1979).

Philosophers of mathematics, mathematicians, and historians of mathematics have thought long and written much on the nature of mathematical objects. To resolve the difficulties they have faced we must first recognize that the higher mathematics as we know it in the contemporary international community of mathematics is a social and cultural development. Second we must realize the nature and consequences of mathematics as a social system subject to professionalization and bureaucratization and the accompanying possibilities for systemic closure and autonomy. In this way we can come to see mathematical objects as the *activities of mathematicians*. Mathematical objects embody their social and cultural histories inside and across the boundaries of variously institutionalized mathematical communities. As the social system of mathematics becomes more institutionally autonomous, mathematical objects increasingly become the resources and tools for manufacturing more mathematical objects. Mathematicians become progressively more insulated from resources and tools from outside of mathematics per se. Organizationally, strong links are developed and sustained across generations in highly reflexive and competitive form. Teacher-pupil lineages and competition among lineages in Western mathematics have historically been strong, and become stronger still in the modern move to greater institutional autonomy (on the general theory of lineages and social networks as the generators of ideas and motors of creativity, see Collins 1998).

Mathematics for its own sake, pure mathematics, and mathematics as an autonomous practice all have the same sociological meaning: that mathematicians communicate more often and more intensely with each other than with outsiders. It does not mean that mathematics or mathematicians are independent of social and cultural causes or that they are context independent. Mathematics is an experimental science that develops in a complex environment of multiple inter- and intra-systemic causes. This is the strong sociological view of mathematics originally adumbrated in the contributions of Emile Durkheim and Oswald Spengler. This is the view of mathematics that has become increasingly prominent in the history, sociology, and philosophy of mathematics and among mathematicians; mathematics is not a transcendent Platonic construct but a matter of practice, experience, and shared meaning in the philosophy of mathematics, the philosophy of mathematics education, and among reflective mathematicians.

The Western pattern of institutional autonomy and competitive self-awareness began to crystallize in the 1600s and escalated in the following centuries. As a consequence, we saw an escalation in the degree of reflexivity and inventiveness in mathematics and the emergence of the hyper-reflexive concerns of twentieth-century foundations research.

Conclusion

> All thought, in its early stages, begins as action.
> The actions which you [King Arthur] have been
> wading through have been ideas, clumsy ones of
> course, but they had to be established as a foundation
> before we could begin to think in earnest.
>
> T.H. White's *Merlyn the Magician*

> What, 'In the beginning was the Word?' Absurd.
> Then maybe it should say 'In the beginning was the
> Mind?' Or better '... there was Force?'
> Yet something warns me as I grasp the pen,
> That my translation must be changed again.
> The spirit helps me. Now it is exact.
> I write: 'In the beginning was the Act.'
>
> Goethe's *Faust*

The sociology of mathematics situates mathematics socially, culturally, and historically. The casual student of this field (not to mention some professional students of mathematics) is urged to recognize that this perspective is applied to both the *external* societal level institutional and organizational conditions which contain number work and mathematical practices, as well as to the *internal* organization of the mathematical practices within communities and social networks of mathematicians. On the one hand the notion of "internal and external factors" is an analytic device (Stingl 2011, forthcoming, see also: Shapin 1992),

On the other is an ideological one that supports certain ideas about purity and causal factors. The Spenglerian idea of mathematics as a world-view is not, in the end, compatible with internal-external analysis.

The mathematics of any particular time embodies its own social history. This process becomes increasingly intense as and to the extent that mathematical activity becomes and remains more clearly differentiated from other social activities and more autonomous. The power of mathematics is not in formal relations between meaningless symbols, nor a transcendental revelation. The implied criticism of philosophy here is based on the understanding that classical philosophy has been a kind of generalized Platonism and to that extent it has obstructed efforts to ground mathematics (along with science, technology, logic, and god) in social life. There is now an alternate philosophy we can turn to that focuses on practice. In this chapter the focus has been on mathematics and sociology as distinct viable disciplines, and philosophy has for the most part been viewed as obstructing rather than facilitating our understanding of mathematics. In the larger context of the project of Science/Craft, we begin here to see some of the outlines of how these disciplines are merging in an interdisciplinary environment leading we conjecture to a neo-natural philosophy.

The idea that our ideas are grounded in our everyday reality has found its way into literature and philosophy even as philosophers and other intellectuals have continued to defend one or another form of Platonism. In the opening epigraphs to this conclusion, T.H. White has Merlyn the Magician proclaim that "all thought begins in action," echoing the sentiment expressed by Gauss through Faust who says that everything begins with "the Act."

The recurring theme in this chapter is that we cannot understand mathematics by focusing on technical talk—philosophical or mathematical—about mathematics. That classical approach isolates mathematics from practice and tends to spiritualize mathematics as an inner-directed technology. Science and technology studies has helped us to understand technical talk as social talk. Mathematics is not, to borrow anthropologist Clifford Geertz's (1983: 98) vocabulary in his analysis of art, "concatenations of pure form," "parades of syntactic variations," or sets of "structural transformations." Geertz has the same objective in this analysis that we have here, and that is to critically replace a context free view of a social activity with a social constructionist view. The question "What is mathematics?" in this sense reveals it to be a sensibility, a collective formation, a worldview, a form of life. Mathematics and mathematical objects are best understood in terms of a natural history, or an ethnography of a cultural system. To answer this question requires that we immerse ourselves in the social worlds of mathematics ethnographically. It is "math worlds," social networks of cooperating and conflicting human beings, that produce mathematics, not individual mathematicians or the minds or brains of mathematicians.

Math worlds as social worlds are grounded in epistemological vernaculars (Stingl 2011a); they are figured worlds (Holland 1998). Cognition and action in math worlds are shaped by a politically infused culture consisting of a constellation of ideas and meaning laden physical forms (Mukerji 2010).

The short hand phrase, "the social construction of mathematics," refers to the social fact that mathematics, mathematical objects, and mathematicians themselves are manufactured out of the social ecology of everyday life and its affordances (i.e., the locally available social, material, and symbolic interpersonally meaningful resources). Social constructionism is not a philosophical statement, claim, or concept. It is in fact the fundamental theorem of a sociology broadly conceived, a sociology being transformed by the interdisciplinary movements of our era. This emerging form of sociology has the potential to become genuinely postcolonial (Go, forthcoming), and even to become an act of "epistemic disobedience" (Mignolo 2012, Stingl 2012e).

The social construction theorem entails that our thoughts and behaviors are not simple or otherwise straightforward products of revelation, the soul (both of which we recognize as being historical ideas but which we eliminate a fortiori because they are imagined causal roots), genetics, biology, mind, or brain. All of our cultural productions come out of our social interactions in the context of sets of locally available material and symbolic resources embedded in social ecologies. It is the case that our social interactions unfold on the scaffoldings and substrata of our biologies but they take place on a social level of reality. The idea of the social is at once transparent (we are, after all, social beings through and through and see and experience relationships all around us all the time) and invisible. It is widely understood that we long ago despiritualized the physical and natural levels of reality; the discovery of the social level is still not widely known or appreciated in a robust sense. For the form of interdisciplinarity we champion in this book to be sustainable, it must support the continuing emergence of a robust social science even as it brings the social sciences into closer and closer association with our biology (cf. Clark 2010, Noë 2010, and Reyna 2007).[14]

Mathematics is a product of human life and therefore a social creation, rooted in our material earthbound reality and human labor. How is it, then, that Platonic ideas come to impact mathematical work? First, we must take into account the effect of "universals." All humans share certain qualities and facts of life around which cultural differences are woven. And all human beings live under environmental conditions that however varied are reflections of the physics, biology, and chemistry of life, culture, space, and time. The universals that follow from these conditions are the foundations for the universals of mathematics. Everywhere in everyday life, if we place two apples next to two apples the phenomenological result will be four apples. The multiplication of experiences like this with different objects leads to the generalization that $2 + 2 = 4$. But that symbolic generalization is culturally glossed and means something very different in Plato, Leibniz, Peano, Russell and Whitehead. The everyday world of picking and adding apples to a barrel is in all-important respects the same for Plato and

_____

14   And see Barnard 1938 for a classical adumbration of this idea.

for Russell and Whitehead. However, the scope and depth of human experience has changed dramatically in the millennia that separates their lives and so has the social world of mathematics. The is number work, like adding up apples, that is constant across the centuries and in important respects across cultures (taking into account at the very least translation caveats). Culture always intrudes in our mathematics; but while there is no Mathematik, there are ways for us to translate and communicate commensurable number and math work across cultures.

To understand mathematics as a social construction we have to be ethnographers of mathematics. How do people become mathematicians? What are the conditions under which they are educated, professionalized, and enter into world worlds? What are the conditions like at their work sites? What are the materials they use, what is in their tool box, how does mathematical production unfold? To answer such questions is to describe the culture of mathematics. An ethnography of mathematics would describe the mathematicians tools, techniques, and products (material culture), patterns of organization—social networks and structures, patterns of social interaction, rituals, norms, values, ideas, concepts, theories, and beliefs (social culture), and the reservoir of past and present symbolic resources that they manipulate in order to manufacture equations, theorems, proofs, and so on (symbolic, conceptual culture. Ethnographies and historical sociologies of mathematics must, to be complete, situate mathematics cultures in their wider social, cultural, and global settings. They must also attend to issues of power, class, gender, ethnicity, and status inside and outside more or less well-defined mathematical communities. Olaf Breidbach has come to characterize this recently as a project in "radical historicization" (*Radikale Historisierung* (2011)). For two outstanding example of what this means in practice, see Rosental's (2008) historical sociology of logic and Nye's (1990) sociologically inspired philosophical history of logic.

To say that "mathematics is a social construction" is to say furthermore that the very products of mathematics—mathematical objects—embody the social relations of mathematics. They are not free-standing, culturally or historically independent, Platonic objects. To view a mathematical object is to view a social history of mathematicians at work. It is in this sense that mathematical objects are real. Before there is mathematics, there is number work; before there are professional mathematicians, there are number workers and then mathematicians.

A Hidden Interrogation Revealed

Number workers in households, commercial enterprises, and schools set the agendas in and create the foundations for arithmetic, geometry, and trigonometry in the ancient world and for long periods thereafter. Higher mathematics is the product of increasingly specialized number workers and eventually of variously disciplined professional mathematicians. The historical and ethnographic interrogation of mathematics sociologically inspired unfolds the nature of mathematics and mathematical objects as cultural phenomena. There is a hidden interrogation in the interrogation of mathematics. The interrogation of mathematics by standing Platonism on its head and undermining claims of transcendence and purity leads ultimately and I argue inevitably to interrogating religion and the gods. One only has to consider that the sociologist Emile Durkheim (1912/1995) concludes his remarkable study of the social construction of religion and the gods by arguing that logical concepts are, like religion and the gods, collective representations, this-worldly social constructs. Durkheim manifests the unfolding of the sociological enterprise as an exercise in the rejection of the transcendental. The pursuit of this particular line of inquiry must be left for another time.

What can we say about mathematics and the dynamics of social change? We could, as we saw earlier, adopt a position somewhere along a continuum from Rousseau's pessimism to Ferguson's optimism. To what extent does human history deny our humanity and to what extent does it trumpet humanity's triumphs? This is a fruitless debate unless we can adopt some relevant metrics. Perhaps we could adopt the model of progress introduced earlier in the sections on Lenski and Sahlins and Service. Where and how would mathematics fit into such a model? Perhaps the most reasonable view we can adopt is recognize mathematics as one of the many tools humans have fashioned to help them find their way through the complex webs and tapestries of life From the "best of all possible worlds" perspective, mathematics is clearly one of humanity's most useful as well as awe inspiring tools. It is one of the achievements that Nietzsche would have recognized as part of the fragile reason and sense of freedom humans have purchased at great cost. It is ingenuous and historically irresponsible, however, not to juxtapose this view with the fact that mathematics has more often than not been actualized in the service of state power and ruling elites as "weapons of math destruction."

Every civilization from ancient times to the present has made contributions to the mathematics available to us today. In his editor's introduction to Joseph Needham (2000), Nathan Sivin (2000: 13) discusses Needham's hydraulic and metallurgical metaphors for the history of science. The hydraulic metaphor describes the history of any given science or knowledge system as a process in which the rivers of science in different cultures flowed in the great ocean of modern science (the hydraulic metaphor). The problem with this metaphor is its easy linearity, and its implied identification of "modern science" and "Western science." Needham also uses a metallurgical metaphor when he identifies a "transcurrent point" in the history of any science, the point at which the West overtakes China, and a "fusion point" when the two traditions merge. These metaphors ignore the role of conflict and power in determining the movement of ideas across cultures. For example (Sivin (ed.), 2000: 14), the triumph of Western astronomy in China in the eighteenth century was not example of "fusion" but of the appointment of J.A.S. von Bell as an all-powerful Astronomer-Royal by the Manchu overlords. Western chemistry triumphed in China in the wake of the Opium War peace treaties that gave foreign missionaries control of the educational system in China.

Mathematics and science as world heritage systems have supported liberation, social justice, and contributed to making the world a better place for humans. At the same time, mathematics and science have always been handmaidens to ruling classes and power elites and have contributed to the damage and destruction that these classes and elites had led the world into over the millennia. Choose any metric you wish and the picture will be the same; we are at a crossroads in world history, and this is not just any crossroads. The number of roads has been reduced from many to a few, and maybe to only two. One road that seems to opening leads to an industrial-technological dystopia at best and at worst annihilations on various levels and scales. It may be that our fate as a species and a part of a planetary ecology doomed by natural cosmic forces to annihilation on a scale of millennia is now sealed on a scale of centuries or even years. Our survival, and especially our survival with some decent quality of life distributed across the planet and its life forms, will depend on short- and long-term thinking that is at once broad, deep, and wise. All of our civilizational tools will be needed in perhaps our last chance in the short run to "save" ourselves and our planet.

The very idea of progress has always suffered from the criticisms of dystopians and cynics but the twentieth century, a century marked by wars, holocausts, ecological disasters, and radioactive fallout has finally crushed romantic and idealistic dreams of a better world. The idea of progress has given way to a focus on unintended consequences, precautionary principles, and technology assessments. How out of place are the echoes of the British Labour Party's celebration in the mid 1960s of "the white heat of technology revolution." And yet can we say that we are no longer seduced by the promises of the technology lottery.

In 1957 a celebration was organized in honor of the centennial of Joseph E. Seagram & Sons (purveyors of whiskey and spirits). In a startling juxtaposition of alcohol and reason, a group of distinguished was invited to speculate on "the next hundred years." They were invited to be optimistic in order to stay true to the idea (read "ideology") of science. The scientists did their best to be optimistic, but it wasn't easy. The Nobel Prize winning geneticist Herman J. Muller predicted a rosy future *if* we could avoid war, dictatorship, overpopulation, and fanaticism. The geochemist Harrison Brown started his defense of optimism by saying "If we survive the next century." John Weir, an economist and future president of Wilfrid Laurier University began his speech with "If man survives ...." The most bizarre and ominous opening came in the speech by the infamous rocket scientist, Wernher von Braun: "I believe the intercontinental ballistic missile is actually merely a humble beginning of much greater things to come."

Whatever mathematics has been in history and whatever it is today, it will play a key role in humanity's future efforts, successful or not, realistic or delusional, to realize the 'best of all possible worlds." Mathematics is unquestionably an important tool for formulating and solving problems of the human condition that are amenable to quantification. But it is also a humanistic mode of knowing. We are still in important ways heirs to Plato's view of mathematics as a tool for the guardian-rulers of his "republic." In recalling this particular heritage, it is important to remember that Plato resisted poetry and metaphor but he was a master of both. We owe more, within the framework of this chapter and this book, to Nietzsche's conception of mathematics. He valued the rigor and refinement of mathematics and urged that it be brought to bear on all of our sciences. His position was not that this was the best way to apprehend the world (recall his views on the limits of science) but that this would allow us to become more aware of our humanity and our relationships with each other with the things of the world.

Mathematics is not the key to the universe; it is not the ultimate form that knowledge should try to reach. It is a useful tool, maybe an incredibly useful tool, but only one of the tools we need to reach our hopefully civilized goals.

Chapter 2

The Social Ecology of Brain and Mind[1]

Sal Restivo and Sabrina M. Weiss[2]

Preface

This chapter reflects in part an ongoing dialogue between two of the authors, Restivo and Weiss, with Stingl as their interlocutor. Part I reflects an intellectual path that led Restivo from the sociology of mathematics to the sociology of mind and brain. It could be alternatively titled: Prolegomena to Any Future Sociology of Mind and Brain. The very idea of "mathematical practice," situated within the social turn in philosophy, pedagogy, and didactics of mathematics, implies a theory of mind. In the terms of technoscentific practices it represents a figured world or an epistemological vernacular.

Theories of mind may be the informal folk theories of our everyday lives or the more formal theories of professional students of mind. Folk theories of mind are at present still more influential in neuroscientific, biological, and philosophical approaches to topics ranging from mathematics to AI and robotics than are professional theories. The problem is that whichever theory prevails in any given setting or study, or for any given researcher, it is more likely than not to locate the mind in the brain and not in the person. In Chapter 1, we discussed mathematics. We are now obliged to ask which theory or theories of mind are built into our theories and approaches to mathematical practice? If turning our attention to mathematical practice as opposed to focusing on questions of essentials and foundations is a turn to the social, perhaps we should be alert to the possibility of a social turn in our theories of mind.

The chapter is in three parts. The first part reflects as much as possible the undiluted voice of sociology (Restivo) already, however, modified by decades of interdisciplinary training and research, further modified by the voice of philosophy and biology (Weiss), and expanded by the voice of Stingl. In Part II, Weiss's voice is in the foreground, Restivo's in the background; Stingl again is their interlocutor. Parts I and II are written in the appropriate first person. Part III is more integrated

---

1   Some sections of Part I originally appeared in *Red, Black, and Objective: Science, Sociology, and Anarchism* (Farnham: Ashgate 2011). Some revisions to the original text have been introduced here. Part I then is a second iteration of the material that originally appeared in Restivo (2011).

2   This chapter, in three parts, represents a conversation between Restivo and Weiss; therefore, Parts I and III are primarily in Restivo's voice while Part II is in Weiss's.

and is written in the tense of the collaborative "we." The "I" of the narrator in Parts I and II are at once a subjective sign of authorship but also an illusion that masks the social nature of the "I." The "I" is, to recall Nietzsche, a grammatical illusion.

Our interests in brain and mind have followed different trajectories. Restivo's interest in the brain began about the time that President George H.W. Bush proclaimed the 1990s "The Decade of the Brain." Weiss's interest begins shortly after meeting Restivo in late 2010. We cemented our mutual interests while attending a conference at Oxford in December of that year on the brain as a focus of scientific and public interest and the emergence of a brain industry. During this period of roughly two decades, the attention space devoted to brain research had been steadily increasing. This movement shows no signs of slowing down, and the European Commission is planning to announce a new "Decade of the Brain" in 2014 and other nations are poised to make similar commitments.

Our objective in Part II is to offer a snapshot of our effort to develop socially relevant models of embodied and extended mind. Part II, then, is rooted in a dialogue aimed at identifying, mobilizing, and melding our perspectives. In this dialogue, Restivo plays the defender of a sociological territory that can sometimes border on sociologism, while Weiss advocates a more classical notion of the role of the philosopher tutored in biology. While in real life our perspectives overlap and challenge conventional disciplinary boundaries, at times we find ourselves falling into our traditional disciplinary disguises. As a result of this dialogue, Restivo has been developing a model that incorporates multiple intricately interdependent and interpenetrating levels of sensing, perceiving, and acting from neurons to neural nets up through brains and nervous systems to bodies and beyond to social interactions and interaction ritual chains. Weiss has made important contributions to this model and helped in embedding it in the contexts of social ecologies and the *Umwelt*. A key question is whether this is a sufficiently unifying pattern of intra-action that can be used at all levels to model the emergence of higher-order cognitive phenomena while opening up useful avenues for further research and discourse. Additionally, what types of theoretical understandings of related subjects, like embodiment, systems theory, information theory, and complexity theory can be applied to bolster this model? Clark (2010), Noë (2010), Protevi (2009) and Sun (2012) demonstrate that this perspective is already abroad on the interdisciplinary landscape. The Restivo–Weiss model is sketched in Figure 1 at the end of this chapter.

By seeking to reclaim the term "mind" as a useful and constructive concept that is relevant to current questions and understandings in social cognition, Weiss's objective has been to incorporate theories that will help articulate key portions of this model. In Part II, she starts with a discussions of embodied and extended mind, engages transformation through the "apparatus" (following Barad), proposes and defines "mindscape," then expands this discussion back out to the social level. Weiss's voice is the prevailing one in Part II.

In Part III, we describe the contemporary intellectual context and climate of our discourse in terms of social ecology and the *Umwelt*.

Part 1: What Can a Sociologist Say About Brain and Mind?

Introduction

I am engaging brain and mind as boundary objects at the intersection of sociology, philosophy, neuroscience, and biology. My immediate objective is to add a sociological voice to a literature dominated by psychologists, neuroscientists, biologists, philosophers, cognitive scientists, and literary theorists. Bringing sociology into this already interdisciplinary field promises to change the nature of theories about how the brain works and what the mind is as well as impact the applied mind and brain sciences. Sociologists have already begun in small but increasing numbers to take on the challenging problem of the brain in the wake of a traditional concern with the sociology of mind.

There is a caveat concerning "the sociological voice." Sociology has some more growing to do; it is still building a robust disciplinary profile in the public arena. For this reason, even as we move into an interdisciplinary landscape, it is important to take time out along the way to practice a little disciplinary imperialism. The Copernican sociology revolution still needs to undergo its Newtonian and Einsteinian revolutions, though arguably Randall Collins has already made the Newtonian turn. In any case, there will be for some time a tension between the Durkheimian imperative and the interdisciplinary imperative. The fact of the matter is that sociology is extremely vulnerable to disciplinary co-optation. Neuroscientists have been getting credit for bringing sociological insights into the public arena, and biologists have been claiming jurisdiction over the subject matter of society. A recent book (Sun 2012) has the title "Grounding Social Science in Cognitive Science." Imagine a book titled "Grounding Physics in Cognitive Science," (ignoring here flawed efforts in reductionism) or grounding physics in anything other than physics. Physics has a robust disciplinary profile that doesn't tolerate this kind of trespassing. Sociology does indeed have a robust disciplinary profile but one that for various reasons we can explain as sociologists of knowledge is overshadowed by the more visible profile of the low tradition in sociology. On the one hand, sociology needs protection and publicity; on the other hand, the incursions of the neuro-, biological, and cognitive sciences on the grounds of sociology can be viewed as a positive move on the lively interdisciplinary landscape of contemporary scholarship.

This book as a whole also exposes the machinations of a collaboration and our efforts to bridge the divide between sociology and philosophy and explore the sense of a philosophical sociology or a sociological philosophy. The rationale for this objective is the increasing recognition that the problems we humans face today as a biological and cultural species have outrun the problem solving capacities of our traditional disciplines and of the very idea of a discipline as a problem solving tool. We make the additional assumption that we cannot meet our objectives in this era of hybridizing disciplines, categories, and classifications without exploring the possibility of mutually grounding social and biological mechanisms. To the extent that we can realize our objectives, we should be able to put ourselves on the pathway to a new understanding of the brain as a social-biological hybrid and to reconstruct mind as embedded and extended (building on the classical contributions of Durkheim, Mead, and Vygotsky and the more recent contributions of R. Collins, S. Restivo, Noë, and Clark). We will in the end have to leave sociology, biology, and philosophy behind and remake the landscape of sociocultural theory in terms of a theory of social ecologies and the perspective of a neo-natural philosophy (or better, natural inquiry). We see this as a direction rather than a goal we can achieve in the limited confines of this book.

An immediate and transparent rationale for this project can be found in "invitations" to social scientists from leading figures in the philosophy of mind, neuroscience, and cognitive science. The philosopher John Searle (1992: 128) writes that he is convinced of the special role "other people" play in structuring consciousness but he doesn't know how to demonstrate or analyze this social element. The neuroscientist Antonio Damasio (1994: 260) understands that we must take social and cultural content into account when studying the brain but he finds the task "daunting." And the cognitivist mathematician and computer scientist Stan Franklin (1995: 10) points out that ignoring cultural factors in mind studies is the "major deficiency" in this field. He suggests that anthropology and sociology should perhaps share a corner with cognitive psychology in the matrix of mind and brain studies. And even in artificial intelligence research, projects like Rodney Brooks' robotics project COG, Cynthia Breazeal's sociable robot KISMET, and the embodiment perspective (Torrance 1994, Varela, Thompson, and Rosch 1991) are opening pathways for social and cultural studies of mind and brain. The point here is that it is just at that point where philosophers, neuroscientists, and their immediate colleagues give up on grasping how to mobilize society, culture, and the self once they've recognized their significance that sociologists can step in and help. And as for those non-social scientists who are adopting and co-opting social and cultural theories, methods, and perspectives, they inevitably fall short of what social scientists have already achieved and can achieve.

Let's pause to assess where we stand today in brain studies and the sociology of the brain. In December 2010, an interdisciplinary conference on the brain in contemporary society was held at Oxford University under the auspices of and on the site of the Said Business School: *Neurosociety ... what is it with the brain these days?* Restivo and Weiss were participants in this conference. The call for papers begins by referencing Patricia Churchland's book, *Neurophilosophy,* published in 1986. In the wake of Churchland's usage, a variety of "neuro-" fields of inquiry emerged. Sociology was not a prominent generator of neuro- fields. Philosophy was described in the call as "the handmaiden of neuroscience," paving the way to a science of the brain over the obstacles of cultural prejudices and superstitions. The ideological scaffolding of the conference organizers, revealing the prejudices and superstitions of a large segment of the intellectual community, grossly overstated the ability of brain imaging techniques to "allocate mental functions to precise cortical areas" and identify the neural pathways of information flows and decision making. These developments, the call claimed, have undermined our belief in free will, revised our understanding of reason, and made us suspicious of traditional "distinctions of kind" which distinguished humans, other animals, and machines. They also brought into question traditional ideas about selfhood and personhood. The conference organizers, oblivious to the questions raised by the development of the social sciences over the last 200 years, seemed to think that now, on account of brain studies—and brain imaging in particular—traditional ideas of self, free will, and consciousness had been revealed to be features of "folk psychology" and were finally giving way to science per se. They thought to ask whether there might be alternative philosophical interpretations for developments in the brain sciences, again oblivious to the findings and discoveries of the social sciences and how they might bear on these developments. This is not to claim that social science had no voice at all here, only that it was a weak voice at best and always subsidiary to other disciplines. The organizers did think to raise questions about the implications of the brain sciences for the moral foundations of personhood, law, religion, politics, and the arts. The stated purpose of the conference was to discuss and debate these issues in the context of a decidedly humanistic framework tied to the post-Enlightenment university and the emergence of a "neuroculture."

In the following year, Restivo and Weiss presented on a panel organized at the annual meeting of the American Sociological Association on *"Neuro": Interventions, Entanglements, Futures*. The call for papers for this panel is an interesting contrast to the Oxford call. Instead of Churchland's philosophy monograph, this call begins by referencing President George H.W. Bush's 1990 announcement of NIH's "Decade of the Brain." The focus of the ASA panel was to be on understanding the brain as a "social/cultural/material object." The concern with the emergence of the brain sciences onto to the scientific and general public arenas of discourse was as prominent in the ASA call as in the Oxford call. The questions to be addressed, on the other hand, were noticeably different though there were overlaps. How are the brain sciences being "created, sustained, contexted, and publicized;" what issues are abroad in terms of funding, brain research, and lab practices; what is the status of *critical* (our emphasis) ethnographies of neuroscience?; and, to quote directly from the call, what is the status of

> narrative and literary approaches to neurology, neuroscience, and subjectivity; technologies in/of the brain, including neuro-prosthetics, neuro-disabilities, and neuro-justice; scientific and public understandings of the brain-body, or mind-body interface; neuro-politics such as neuro-feminism, neuro-diversity, neuro-racism, neuro-utopias; ethical and policy issues; consciousness; and the brain and inequality?

The two calls for papers illustrate through their differences two of the major perspectives on the brain sciences we see today. In one (Oxford), there is an atmosphere of virtually uncritical awe at the supposed power of the brain sciences and the centrality of the brain in the causal nexus underlying behavior and consciousness. At Oxford, one could say society and culture themselves were being neuro-ized. In the other (ASA), the approach is at once broader in scope, more critical, and alive with issues of social justice.

In 1943, McCulloch and Pitts helped set the modern agenda for an immanentist approach to brain and mind. They claimed that logic was the appropriate disciplinary approach to studying the brain and mentality. This idea had already been rejected by Emile Durkheim early in the twentieth century. He denied on social and cultural grounds that concepts such as Aristotle's categorical imperatives and Kant's categories are either (a) logically prior to experience, immanent in the human mind, or otherwise a priori; or (b) crafted by individuals (literally). We can loosely trace the rejection of transcendence and immanence back to the ancient Greeks. For example, Cicero said of Socrates that he brought philosophy down to earth. Perhaps (and arguably) Protagoras initiates the emergence of the social sciences when he announces that "man is the measure of all things." The contest between idealists and realists is a perennial one. Marx had to stand Hegel on his head, the Marxist mathematician Dirk Struik had to challenge his Platonist colleagues, and contemporary sociologists of knowledge still struggle to overcome Platonist myths and purist ideologies in science and mathematics.

The type of fully social theory of mind one finds in the work of Randall Collins (1998) and Restivo (1999) builds most directly on the contributions of Durkheim (1995/1912) and G.H. Mead (1934). The most prominent social theories of mind in contemporary research draw more on the works of L. Vygotsky than his contemporary Mead. Vygotsky is not as radically sociological as Mead, and thus appeals to the more psychologically minded researchers in AI, robotics, and social cognition. The Meadian approach will eventually have to prevail over Vygotsky if we are going to construct a robustly sociological theory of mind. And it is such a theory that will have to guide social robotics in the future if that field intends to produce interactive robots that will fit (to the degree possible) efficiently and effectively into everyday social life. Given that sociology has been silent and silenced in mind and brain discourse, a certain amount of disciplinary imperialism is required upfront to bring in the sociological voice. But our objective is in the end to give voice to a "new" sociology that is interdisciplinary and prominently interdisciplinary with respect to biology and philosophy.

It is also important to register in these early moments that sociology has something to say about the brain. Clifford Geertz (1973: 76) has pointed out that the brain is "thoroughly dependent upon cultural resources for its very operation; and those resources are, consequently, not adjuncts to but constituents of, mental activity ...." Indeed, DeVore (Geertz 1973: 68) has argued that primates literally have "social brains." The evidence for this conjecture in humans has been accumulating in recent years along with a breakdown of the brain/mind/body divisions (e.g., Brothers 1997, Pert 1997).

Our understanding of mentalities has been obstructed by some deeply ingrained assumptions about human beings. One is that affect and cognition are separate and separated phenomena. This division is breaking down (e.g., Damasio 1994, Gordon 1985, Pert 1997, Zajonc 1980, 1984), and will have to be eliminated as part of the process of constructing a sociology of mind. Another assumption is that learning and cognition can be decontextualized. By contrast, social scientists have increasingly argued that learning and cognition are linked to specific settings and contexts, that is, they are indexical. Their long-term efficacies are in fact dependent on contextual recurrence, contextual continuity, and recursive contextualizing. These processes help explain generalization without recourse to epistemological mysteries or philosophical conundrums.

As we move from one sub-environment of our everyday lives to another we do not encounter wild differences and divergences. The spatial structures of our homes, our schools, our stores, our hospitals, and our streets and buildings are alike in size, shape, and context, enough so that we are not constantly required to invent or call on new strategies to negotiate them as we travel through the world. Even as we travel *across* the world, certainly in the more industrialized centers, we do not expect surprises that will stop us in our tracks with their strangeness. Recurrence, continuities, and contexts in our spatial and temporal environments reduce surprises.

We live our lives by moving from home or school to home or school, from our home to our neighbor's home, from the schools we attended to the schools our children attend. Contexts repeat, imitate, suggest, overlap, impose and re-impose themselves, shadow and mirror each other, and are linked through simple and complex feedback loops. This is the structural and informational basis for the continuities in our sense of self, our memories, our thoughts. Many of the mysteries of the paranormal and our everyday experience of déjà vu can be explained by attending to recurrence, continuities, and context.

Recent discoveries about brain plasticity and neuronal regeneration have opened doors for social and cultural influences on the brain by challenging old dogmas about the brain as a free-standing computer-like stabilized machine. Neuroscience, while still haunted by the brain in a vat ideology, is increasingly treating the brain more like a person—acting, subjective and embodied. But we still need some actual persons to put around that brain. Contemporary psychiatric and neuroscience research conclusively demonstrates that mind/brain responds to biological and social forces and is jointly constructed by both. The cytoarchitectonics of the cerebral cortex is sculpted by inputs from the social environment. Socialization shapes the essential human attributes of our species. The dynamical multi-leveled autopoietic-allopoietic approach, as an overarching description or maturation of biological and social systems using a common set of variables, can be a basis for the study of the brain as an aggregation system.

The social and cultural understanding of mentalities has been obstructed by some deeply ingrained assumptions about human beings. Our efforts in developing a sociology of mind and brain have co-evolved with and intersected the works of others at the intersection of the social, life, and neurological sciences. While we will be working toward an interdisciplinary integration of social science, biological, and neuroscience paradigms, Part I is about how our work stands apart by virtue of its emphasis on sociological thinking.

Theories of Mind

Traditionally, theories of mind, primarily coming out of philosophy and psychology, have been asocial. These theories include Hegelian mentalism, idealism, materialism, dualism, various forms of monism, and variations on these themes including Cartesian, bundle, interactionist, parallelist, behaviorist, logical behaviorist, functionalist, phenomenological, central state or identity theories, and various attribute theories (Armstrong 1968, Priest 1991). One of the most intriguing aspects of how philosophers and psychologists think about minds is the effort to explore the nature of the human mind by imaginings about brains in vats, armadillo minds, thinking bats, Martian brains, philosophical zombies, and so on. No wonder we can't find social human beings anywhere in these theories!

Traditional and prevailing approaches to mind and mentality in general center on the brain. Mentality is viewed as either caused by or identical with brain processes. Given this perspective, John Searle (1984:19) could argue that "pains and all other mental phenomena are just features of the brain (and perhaps the rest of the central nervous system)." But Durkheim's analysis of different degrees of social solidarity and the social construction of individuality suggests a culturological conjecture on pain: the extent to which a person feels pain depends in part on the kind of culture s/he is a product of and participant in, and in particular the nature and levels of social solidarity in the social groups s/he belongs to. Furthermore, the symbolism of the pain experience in its cultural context is also a determinant of felt pain. "Pain" has a context of use, a grammar. Such a conjecture was indeed already formulated by in *the Genealogy of Morals* Nietzsche (1956/1887: 199–200). Wittgenstein's (1953) writings on pain in his *Philosophical Investigations* provide additional ingredients for a social theory of mind based on the role of language in our pain narratives. But Searle, while he invokes the social, does not know how to mobilize it theoretically, and so argues that consciousness is caused by brain processes. We will see as we proceed why this claim that has seemed so reasonable for so long must be reconsidered in light of what we know about the relationship between social life and consciousness, and what we are learning about social life and the brain.

Cognitive psychologists tend to view the mind as a set of mental representations. These representations are then posited to be causes behind an individual's ability to "plan, remember and respond flexibly to the environment" (Byrne 1991: 46). Cognitivists also have a tendency to equate cognition and consciousness. But Nietzsche long ago had the insight that consciousness is a social phenomenon. He was one of a number of classical social theorists who had pioneering insights into the social nature of mentality.

We can approach the history of discourse on mind in terms of (1) the conflict between rationalists (intellectual descendants of Descartes and Leibniz) and empiricists (followers of Locke, Berkeley, and Hume); (2) the behaviorist challenge to the radical empiricists by Watson and others, and the challenge in turn to the behaviorists by the ethologists (Lorenz, Tinbergen, and von Frisch); and (3) the Kantian counterpoint to empiricism, represented in our own time, for example, by Jerry Fodor's (1983) conception of the mind as an entity possessing organizing capacities and an innate "language of thought."

Why is it we "locate" mind, thinking, and consciousness "inside" heads? Certainly in the figurative world of the "Western colonial matrix of power" (Mignolo 2012), mentalities and the emotions have been associated with the brain and/or the heart since at least the time of the ancient Greeks. More recently, localizationalist physicians and neuroscientists have reinforced the idea that mentalities are "in the head" (Star 1989). On the other hand, in sociological perspective, mentalities are not produced out of or in states of consciousness; they are not products, certainly not simple products, of the evolution of the brain and brain states. Rather, they are by-products or correlates of social interactions and social situations. This implies that the "unconscious" and the "subconscious" are misnomers for the generative power of social life for our mentalities—and our emotions. There is no more an unconscious than there is a god, but there are cultural mechanisms for translation and transference that point us to referents that do not exist. The thesis here is that social activities are translated into primitive thought "acts," and must meet some filter test in order to pass through into our awareness (cf. Wertsch 1991: 26–7; and see Bakhtin 1981, 1986 and Vygotsky 1978, 1986). In the wake of the pioneering contributions of Lotze, Janet, and William James (van der Veer and Valsiner 2000), Vygotsky, Bakhtin, and G.H. Mead should be recognized as the inventors of the modern social theory of mind. Wertsch (1991: 14) stresses that mind is mediated action, and that the resources or devices of mediation are semiotic. Mind, he argues, is socially distributed mediated action.

Getting to the Beginning of our Story

In 1939, C. Wright Mills (1963) argued that the sociological materials relevant to an understanding of mind had not been exploited by sociologists. Mills had in mind in particular the work of the American philosopher and social theorist, George Herbert Mead. Fifty years later, Randall Collins (1988) could still write that Mead's writings on the sociology of mind were underdeveloped and unexploited. Today, 25 years after Collins's remarks, we are justified in making the same claim. Indeed, the failure of sociologists to pick up the track of Mead's social theory of mind was underscored by the publication of a social theory of mind guided by Mead's work written not by a sociologist but by a neuroscientist/ psychiatrist (Brothers 1997).

Resources for a Sociological Theory of Mind

The basic resources available for constructing a sociological theory of mind include but are not limited to the following: the concepts of *collective representations* (Durkheim) and *generalized other* (Mead); Goffman's (1974) *theory of frames* (cf. Wertsch 1991 on *recontextualization*); the literature on *culture and thought* (Cole and Means 1981, Goody 1977, Levi-Strauss 1966, Levy-Bruhl 1985/1926); studies of the evolution of *human language and its social context* (Caporael et al., 1989), and studies *on the relationship between social relations and rule-governed systems such as language* (Caporael 1990: 10–11). Researchers in artificial intelligence have been increasingly incorporating into their work the idea that AI *machines have to be programmed with "cultures"* (e.g., Keesing 1987: 381; Normal and Rumelhart 1975). By the 1990s, these and related ideas had coalesced into efforts to build affective computers and sociable robots (e.g., Breazeal 2002, Picard 1997, Restivo 2001).

Mind and Brain: What Are We Thinking?

Headlines assigning agency to the brain have been a prominent feature of our media landscape for almost two decades, with the brain being credited for everything from creating God and generating our personalities and emotions to making moral decisions and playing chess. The brains of celebrated personalities from Lenin to Einstein have been analyzed for clues to political and scientific genius. Philosophers and neuroscientists, however, have begun to acknowledge that 'the other' and 'culture' must play a role in what the brain 'does.' Some, like philosopher John Searle and neuroscientist Antonio Damasio, have been openly baffled by this challenge. Others, like psychiatrist and neuroscientist Leslie Brothers, have addressed this challenge brilliantly. Drawing on the ideas of neuroscientists such as Brothers and anthropologist Clifford Geertz's pioneering speculations on brain, mind, and culture, it is now possible to construct a foundation on which to build a social theory of the brain. Such a theory has dramatic implications for understanding what the brain is and how it works. Even more significant are the practical implications of a social understanding of the brain for dealing with mental illnesses and strokes, and with the brain in health and illness more generally. The approach I adopt here also leads to a new way to understand socialization. The very idea of a sociology of the brain is a radical departure within sociology even given the era of science studies that has given us sociologies of mathematics, logic, and god. On the other hand, the sociology of mind has a track record in sociology that needs to be more widely recognized.

According to Astington (1996: 184), Gopnik (1996) claims there are only "three games in town" when it comes to theory of mind: theory-theory, simulation theory, and modularity theory. But Gopnik (1996: 169, 182) distinguishes "theory-formation theory" from "theory-theory." Nonetheless, the psychologism in these theories fits the individualist bias we find in work ranging from research on children's' theories of mind to social robotics. There is another game in town, however, and it goes with the sociological imagination. The alternative to children deriving their theories of mind from their direct experiences of such states, of developing such theories the ways scientists supposedly derive their theories, or of giving rise to them innately as they mature is an enculturation theory.

The prevailing theories of theory of mind emphasize development within the individual. From a sociological or anthropological perspective, theory of mind and mind itself are cultural inventions (Astington 1996: 188). Social construction of mind has not been ignored, but it has not been as centrally represented in either mind studies or social robotics. The reason is a problem in the sociology of knowledge rooted in the fallacy of introspective transparency and the myth of the technological sterility of sociology. These are not failures of sociology but rather failures of the sociological imagination in philosophy, neuroscience, and robot science and engineering. Similar problems accrue to educational theories in mathematics to the extent that they are grounded in traditional psychology and philosophy.

The Social Mind

The sociology of mind and thinking has a long and distinguished pedigree, yet it has until recently been virtually invisible in contemporary theories of mind (Valsiner and Van der Veer 2000). A renewed interest in mind, brain, consciousness and thinking is evident in the steady stream of books, articles, lectures, news stories, and television programs crossing today's intellectual and lay landscapes. One of the main features of this media storm is that one can see some evidence of a sociological orientation emerging, albeit timidly and fearfully, out of the shadows. It is no coincidence that we see at the same time increasing interrogations of god and religion. Mind, math, and god problems are linked to the extent that they have traditionally been treated as if they arose outside the constraints of human social activities.

An archaeology of these developments would reveal a "journey to the social" across the entire landscape of intellectual labor. Those undertaking this journey who are not sociologists or anthropologists (or more generally, social scientists) stop short of their mark or otherwise abort the trip. This is, indeed, a much more treacherous journey than the westerners' journeys to the east which have captivated (and captured) so many western seekers. But the very fact of the journey to the social reveals the emergence of a new discursive formation, a new *episteme*.

This *episteme* is new in the sense of a birth or an originating activity, but absolutely new in the scope of its impact. Beginning in the 1840s, the west entered the age of the social, an era of worldview changes that will carry well into the twenty-first century and likely beyond before it begins to embody itself in the everyday ecologies and technologies of intellectual life in new global configurations. In this process, what was Western and European about the social will get permeated by the Other and transformed into a worldview that is less ethnocentric.

As co-authors we have arrived at this collaboration by different routes and intellectual genealogies. What unites and divides us are equally important. What we share is the realization that "the biological" and "the social" cannot be neatly labeled and packaged into black boxes that do not speak to each other. We reject reducing human social life to biology and at the same time imprisoning it in the grip of an imperialist sociology. The Durkheimian sociologist, Restivo, the classic natural philosopher and nomadic political theorist, Stingl, and the bioethicist–interdisciplinary philosopher Weiss (bracketing their ethno-gendered standpoints) agree that humans must be understood in a *naturecultures* perspective. We share the road to social and bio-cultural studies of mentality with artificial intelligence researchers and engineers (including notably Rodney Brooks and Cynthia Breazeal), embodiment theorists of mind such as Torrrance (1994), and embodied action theorists of cognition (notably Varela, Thompson, and Rosch 1991).

The Sociology of Mind

In consideration of the conclusions reached in Chapter 1, the idea that the mind is a social construction is crucial to reforming our understanding of mathematics education. In particular, Restivo has been concerned over a major part of his career with bringing mathematics down to earth. To bring mathematics out of the platonic clouds, out of transcendental realms, is equivalent to negating the idea of "pure mind."

When, and to the extent that, mathematics becomes a functionally differentiated, institutionally autonomous social activity in any given social formation, it will begin to generate mathematics out of mathematics. The vulgar notion that "mathematics causes mathematics" (pure mathematics) arises out of a failure to (and to be able to) recognize that in a generationally extended mathematical community (or social network of mathematicians) using the results of earlier generations of mathematical workers and mathematicians as the (material) resources for their mathematical labors. Systematization, rationalization, and generalized ("abstraction") in mathematics are dependent on organizing mathematical workers in a certain way. In general, this means specialized networks and sustained generational continuity. The widely recognized significance of iteration as a factor in mathematical development and creativity is dependent on the social iteration produced by generational continuity.

For centuries, it has seemed obvious that the study of mind should be under the jurisdiction of philosophers and psychologists (in their pre-modern as well as modern guises). As the matrix of mind studies became increasingly interdisciplinary in the latter part of the last century, sociology and anthropology were notably left out in the cold. It may be that these are the only modes of inquiry that have any hope of making sense out of the chaos of claims about mind, consciousness, and even god and soul coming out of contemporary physics, astronomy, biology, artificial intelligence, cognitive science, and the neurosciences.

Sociologists like Randall Collins and Restivo, Norbert Wiley and Margaret Archer begin their efforts in the sociology of mind by making a simplifying assumption—that thinking is internal conversation. This poses an immediate problem. That is, given everything written so far, and given Wittgenstein's writings on mind and thinking, we do not want to claim that thinking (as conversation, for example) is something that happens inside heads or brains. There are efforts abroad to develop an explanation of cognition as embodied action. A theory of embodied action that is properly sociological dissolves the inner/outer dilemma and the chicken/egg problem. The chicken point of view is that there is a world "out there" with pre-given properties. These exist before and independently of the images they cast on the cognitive system. The role of cognition is to recover the external properties appropriately and accurately (realism). From an egg perspective, we project our own world, and "reality" is a reflection of internal cognitive laws (idealism). But a theory of embodied action explains cognition/ mentality in terms that depend on having a body with a variety of sensorimotor capacities embedded in more encompassing biological, psychological, and cultural contexts (Varela, Thompson, and Rosch 1992). Cognition is lived; sensory and motor processes, perception, and action are not independent; sensory apparatuses are not even independent of each other. This approach promises to dissolve the inner/outer dilemma, and to eliminate representational paradoxes in the theory of mind. Details on how such a perspective bears on our understanding of how we learn mathematics can be found in Stephen Lerman's (1994) work. There is already more than a hint in the embodiment imperative on how to solve the mind/body problem. We do not socialize individuals, persons, or selves as sociology textbooks and theorists are wont to claim. That claim assumes as already extant what socialization aims to create. What then is socialized (assuming that that is still an appropriate term)? Suppose that what is socialized is the brain/central nervous system, with the b/cns in a social ecology, a network of social relationships and interactions? A theory constructed along some such line would eliminate the mind/body problem and make a rapprochement between neuroscience and social science a reasonable expectation.

In order to cement the relationship between embodiment and the sociology of mathematics, for example, consider what we mean by endless counting. Following Restivo, inspired by Nietzsche, Wittgenstein, Rotman and others, we must ask in the first place who is it that will be making the endless marks or speaking the endless numbers. It must be either us or some surrogate. We could only relate to such a surrogate if it was like us in significant ways. If what is significant about us is our embodiment, any counting surrogate would have to be embodied. The alternative is a disembodied agent counting outside of time, space, and materiality. An embodied counter undermines Platonism and absolutism and gives us the grounds for a sociology of mathematics. Numbers then appear as things mortal embodied humans have to make and re-make in a finite context (cf. van Bendegem 1987).

For the moment, we want to focus on the "phenomenology" of a certain kind of thinking experience. We do not mean here what Parsons called the fictive and contemplative philosophy developed in the works of Alfred Schuetz and his disciples. A sociological theory of mind must account, one way or another and sooner or later, for the experience of "inner thought." and it must do so without the assumption or claim that this experience is universal across humans and cultures. Conversation is the prototype for a certain kind and level of thinking, the kind of thinking we, initially at least, have in mind (so to speak) when we set out to construct an artificial intelligence, develop a theory of mind, or think about our own thinking. We must learn to speak out loud before we can think "silently," "in our heads." "External" speech already contains all the crucial elements of thought: significant symbols, capacity to take the stance of one's interlocutor or listener, and the ability to take the role of the other and orient to the generalized other (see the discussion below).

Internal conversations do not necessarily have the same structure as external conversations. Short cuts, shunts, and short-circuits in our thinking are possible when we (especially as adults) are thinking smoothly. We may know almost immediately where a thought is going and whether to pursue it or switch over to another thought-track. Because we can monitor multiple thought-tracks (the dispatcher function), we can rapidly switch between alternatives, elaborations, objections, and conclusions. Thought-tracks and trains of thought connect syntactically and pragmatically in Hesse-type networks (Hesse 1974). And words invoke other words, ideas invoke ideas, concepts invoke concepts (because of similar meanings, sounds, and/or associations). Generally, these switches, invokings, and associations occur smoothly and without the exercise of "will;" and they can produce what Restivo calls thought cascades. If the process is disrupted in any way, however, our attention will shift, the process will slow, and we will proceed with awareness. This contributes to the illusion that we think "willfully."

If we treat thinking as internal conversation, then thinking must be constructed out of past, anticipated, and hypothetical conversations. In other words, what we think is connected to our social networks (including reference groups). Then the greater the attraction to given parts of the network, and the closer we are to given subsidiary networks, the more we will "be motivated" to think the ideas circulating in those parts. The connections among ideas are emotional as well as associative and grammatical. Words, ideas, and images have valences. And consciousness itself is a type of emotion, attentiveness. Normally this attentiveness is very mild and attached to certain sign-relations. The level of attentiveness presumably changes as social situations (real and imagined) change. Only when the smooth and easy inference (or "next move") is blocked, or contradicted by something in the situation, does the emotion erupt into consciousness. So emotional weightings (valences) affect what a person thinks about at a particular time. These ideas are consistent with neuroscientific and sociological research that suggest the existence of a baseline emotional state.

The Generalized Other Revisited

The generalized other is the core concept in George Herbert Mead's social theory of mind. Mead introduced the idea of the generalized other to describe that component of the self constructed out of the variety of messages we receive from the people we come into contact with. The generalized other is the source of our ability to take the roles of others, and also the source of our understanding of the "rules of the game" in everyday interaction. It is the locus of what Freud called the super-ego, which gives us "conscience." and it is the locus of what Restivo has come to call "moralogics." When we reason, generalized others are with us all the way, approving and/or disapproving our every move. We always reason from a standpoint. There are many standpoints, and each is guarded by a generalized other. Operating logically means operating in terms of standard and standardized critical and reasoning apparatuses. Individuals cannot be logical or illogical. They can only be in agreement or disagreement with a community of discourse, an objectivity community, a thought collective. And patrolling standpoints is therefore a moral act. If, then, reasoning is always grounded in a standpoint, there can be no general abstract reasoner, no eternal, universal logic. If, furthermore, patriarchy has constructed Platonism, and relativity theory, and truth-seeking Diogenes and the propagandist Goebbels, the podiums of rationality and objectivity and the arenas of emotion, then there is good reason (from a certain standpoint, now!) to conjecture that mentality or mind is "man-made." Thinking is, therefore, on these principles, gendered. Logic is the morality of the thought collective, and it carries the weight of how gender and power are distributed therein. Again, Restivo not only independently resonates with but has anticipated the post-colonial perspectives in science studies—not because this would be such a difficult feat in and of itself, but because it is actually hard to understand how one could not. We have been western colonial dopes and that requires our attention and an explanation.

From such a (stand)point of view, it is easy to conceive that neither "laws of logic," "laws of form" (Spencer-Brown) nor "laws of thought" (George Boole) are intuitive, innate, or a priori. Generalized others carry socially derived logical systems that restrict, govern, filter, direct, and cue logical speech acts. Inside every word, inside every vocabulary, inside every sentence, and inside every grammar we find discourse communities; logics are language games. It follows that our thoughts, insofar as they draw on the resources of languages, are socially textured.

Goffman's (1974) frame analysis provides another ordering apparatus. And the distinction he attends to between conversational talk and informal talk has an analogy in thought. Just as informal talk holds the individual together across parsing moments and breaks in continuity in social projects, and just as much of what we say in the presence of others is related to creating and sustaining social solidarity, so informal thought is about self-solidarity. Speaking, Goffman points out, "tends to be loosely geared to the world." Talk is looser. Thinking is even looser, and more vulnerable to the processes Goffman calls keying and fabricating. This discourse is indebted to Durkheim, Goffman, Wittgenstein, Randall Collins, and notably Norbert Wiley (see in particular, Wiley 1995: 108).

Now let us think again about moralogics. Mathematics communities are in part crucibles for refining the idea of god through exercises with infinity/ies. The most abstract (generalized) efforts then turn out to be tied more or less explicitly to the god project. Boole's goal was to reduce "systems of problems or equations to the dominion of some central but pervading law." This is not a simple metaphor, for Boole was set on establishing the existence of god and a universal morality. So too Cantor's transfinite numbers are implicated in the search for a proof of the existence of god. (We cannot pursue this further here, but see the appendix on mathematics and god in Restivo 1992).

Where is Thinking?

The introduction to the social construction conjecture should make it easier to understand what we mean when we say that minds and thinking are social constructions. This conception carries with it the notion that thinking is a networked and dialogic (if not even "diaphysical" in its full involvement of communicating whole bodyminds or "psychosomata" [Stingl]) process, a series of social acts rather than something that goes on inside isolated, independent heads and/or brains. This does not mean that heads and brains are dispensable, or that neuroscientists and psychologists have nothing to teach us about minds and thinking. But it is social relations that give rise to consciousness and thinking; the genesis of consciousness and thinking is in society not in the brain. Freestanding brains do not and cannot "become" conscious, and do not and cannot generate consciousness in some sort of evolutionary or developmental "brains in a vat" process. Consciousness, thought, and language cannot be explained or understood independently of the understanding that human beings are fundamentally and profoundly social. More importantly, they are profoundly rhythmic, and it is the coupling of their rhythms in social interaction that produces consciousness, emotions, and communication.

Individualized thoughts must be tied to their social bases if we are to understand their genesis and nature. Communicable thoughts are, by definition, shareable and shared (Durkheim 1961: 485). All concepts are collective representations and collective elaborations—conceived, developed, sustained, and changed through social work in social settings. Indeed, Randall Collins (1998) has shown through detailed comparative historical studies that the configurations and developments of social networks of intellectuals cause particular ideas to come into being and develop or die out. This line of thinking leads to the conclusion that it is social worlds or communities that think and generate ideas and concepts, not individuals. Social worlds do not, of course, literally think in some superorganic sense. But individuals don't think either.

Rather, individuals are vehicles for expressing the thoughts of social worlds or "thought collectives." To put it another way, minds are social structures (Fleck 1979: 39, Gumplowicz 1905). Mentality is not a human invariant. And even vision is an activity and not a neurological event (Davidson and Noble 1989, and see Heelan 1983 on the social construction of perception).

In order to grasp the idea that thinking is radically social, and to keep it from slipping into some spiritual or mystical realm, or becoming an empty philosophical or theological concept, one must keep firmly focused on and fully comprehend the idea that humans are social beings and that the self is a social structure. It is also crucial that we do not project our modern post-literate experience of mentality and mind-body duality on all humans in all times and places. "mind" is not a cultural or human universal (cf., Davidson and Noble 1989, and Olson 1986).

Ritual and Cognition

Cognition arises situationally out of the natural rituals of everyday interactions and conversations. These rituals form a chain, and as we move through this chain, we come across and use more or less successively blends of cultural capital and emotional energies (Collins 1988: 357ff.). The concept of ritual developed in the work of Emile Durkheim can be generalized and conceived as a type of framing (following Goffman 1974). This leads to the idea that the theory of ritual can be developed in terms of the different types of framings and reframings, which constitute our movement through interaction ritual chains. From this perspective, solidarity rituals take place in a social market that is variously stratified. Language is a product of a pervasive natural ritual (words, grammatical structures, speech acts, and framings are collective representations loaded with moral significance). The ingredients of language refer outside conversations, and their sense is their symbolic connection to social solidarity and their histories in interaction ritual chains. All thoughts take place in several modalities—visual, aural, emotional, sensual—simultaneously. Indeed, it is the socially constructed, gendered, cultured body-in-society that thinks, not the individual, or the head, brain, or mind.

We can now reenter once again the realm of Chapter 1 and the sociology of mathematics. But I must stress that if we enter without at least some preliminary comprehension that self and mentality are social, the sociology of mathematics will seem like a voyage through the looking glass—without any of the charm of Lewis Carroll's guidance.

Conclusion: An Exercise in Applied Sociology of Mind

After all of this social theory, a mathematics educator looking for "new horizons" may very well still feel at a loss as to how to translate social constructionism and its entailments into classroom practices. Whatever guidelines we can provide for such a translation will seem to be distant from the immediate classroom concerns of mathematics teachers. This is inevitable given the difficulties of coming to terms with what amounts to a new worldview. Let us see if we can come a little closer to the communication essentials of the mathematics classroom.

Given social constructionism and the sociology of mind (and brain), classroom teachers and students must learn to ask some old questions with an expectation that there will be new answers. What are numbers (and what are all the basic concepts and processes that constitute mathematics?). What is a classroom? What are teachers and students? What is learning? What is truth? What does it mean to reason? What is a proof? The trick here is to see all of these old friends as *institutions*. A number of mathematics educators have made significant progress in coming to terms with this perspective and making it accessible to and applicable within the working lives of mathematics teachers (cf., Burton 1999, Ernest 1994, 1998).

Let us adopt the teacher's point of view as s/he enters the classroom armed with the tools of sociology. To begin with, the room is no longer filled with individual students housing individual, freestanding brains. It is now to be seen, to be experienced, as a collectivity, variously cultured and more or less culturally homogeneous or heterogeneous. No assumptions can be made about shared paradigms, practices, and discourses (Sfard 1994: 248). We want to speak in terms of a focus on practice as opposed to reasoning. But, then again, this is a false dichotomy. Given the social constructionist perspective, it follows that all aspects of mentality are forms of social practice. The everyday significance of this is that mathematics teachers are not engaged in 'teaching' students how to think mathematically. Rather, they are initiating students into different corners of the mathematical community, and on different levels of mathematical practice. Therefore, their first concern should not be with the ontology or epistemology of mathematical objects and ideas, but rather with the practices that give rise to and sustain those objects and ideas in the lives of students.

Sfard (1994: 270) recommends (following, for example, Leibniz's philosophy, not to mention Wittgenstein) an emphasis on using concepts in various contexts rather than trying to get students to grasp concepts immediately in the abstract. It would be fruitful if teachers in all fields recognized that learning in general emerges as we use words in different contexts and not out of brute force definitions and applications. It is not shared definitions that make communication possible but rather common practices in pursuit of shared goals.

There is the suggestion in Sfard (1994: 269–71) that mathematics students might learn more effectively by recapitulating the ways the mathematical community came to collectively grasp concepts and ideas. The historical, social, and cultural contexts cannot be separated out from the substance of mathematical objects, concepts, and ideas. As long as these dimensions of mathematics are considered, "humanistic accessories" and "nonessential distractions" from the curriculum of training, teachers will not be motivated to carry out the requisite curriculum changes. Foundational resources for such changes are readily available in the works of Restivo (1983, 1992) on the calculus, mathematical traditions across cultures, and the networks of mathematics (math worlds), Mackenzie's

(1981) study of the social construction of statistics, and Bloor's (1976) pioneering program for a sociology of mathematics.

Collaborations between social scientists and mathematics teachers would be useful. But this will be counterproductive unless the social scientists are mathematically literate and the mathematicians are sociologically literate. It is still the case that sociological literacy is much more rare than mathematical literacy. Sociological literacy depends on a worldview free of certain fallacies. We conclude with a list of those fallacies, identified by Restivo (see his 2011: 170–75 for the original extended version). These fallacies are the raw materials for a set of theorems about our nature as social beings.

- *the transcendental fallacy (the theologian's fallacy)* is that there is a world or that there are worlds beyond our own—transcendental worlds, supernatural worlds, worlds of souls, spirits and ghosts, gods, devils, and angels, heavens and hells. There are no such worlds. They are symbolic of social categories and classifications in our earthly societies and cultures. There is nothing beyond our material, organic, and social world. Death is final; there is no soul, there is no life after death. It is also possible that the so-called "many worlds interpretation" in quantum mechanics is contaminated by this fallacy as the result of mathegrammatical illusions. The world, the universe, may be more complex than we can know or imagine, but that complexity does not include transcendental or supernatural features.
- *the subscendental fallacy (the logician's fallacy)* is that there are "deep structures" or "immanent structures" that are the locus of explanations for language, thought, and human behavior in general. Such "structures" are as ephemeral and ethereal as transcendental and supernatural worlds. They lead to conceptions of logic, mathematics, and language as "free-standing," "independent," "history, culture, and value free" statements. And they support misguided sociobiological and genetic explanatory strategies.
- *the private worlds fallacy (the philosopher's fallacy)* is that individual human beings harbor intrinsically private experiences. The profoundly social nature of humans, of symbols, and of language argues against intrinsically private experiences (as Wittgenstein, Goffman, and others have amply demonstrated).
- *the internal life fallacy.* When we engage in discourses about surrogate counters, imitators, and artificial creatures that mimic, we need to remind ourselves that we are working in an arena of analogies and metaphors. Such efforts carry a high emotional charge because they take place at the boundaries of our skins. Analogy and generalization, if they can be shown to have constructive scientific outcomes, need not obligate us to embrace identity in, for example, building robots. Robots will not have to have "gut feelings" in the identical sense humans have gut feelings because they are organic machines. Even this "fact" needs to be scrutinized. What we "feel" is given to us by our language, our conversations, our forms of talking.

At the end of the day, feelings may not at all be straightforward matters of bio-electro-chemical processes. Electro-mechanical creatures will turn out to be just as susceptible to an internal life as humans once they have developed language, conversation, and forms of talk. This implies a social life and awareness. Roboticists may already have made some moves in this direction with the development of signal schemas and subsumption-based hormonal control (Arkin 1998: 434ff.). The development of cyborgs and cybrids may make this issue moot.

- *the psychologistic fallacy* is that the human being and/or the human brain is/are free-standing and independent, that they can be studied on their own terms independent of social and cultural contexts and forces. This is also known as the neuroistic error (Brothers 2001). It encompasses the idea that mind and consciousness are brain phenomena. Human beings and human brains are constitutively social. This is the most radical formulation of the response to this fallacy. A more charitable formulation would give disciplinary credibility to neuroscience and cognitive approaches to brain studies. These approaches might produce relevant results in certain contexts. Then there might be fruitful ways to pursue interdisciplinary studies linking the social sciences to the neurosciences.

- *the eternal relevance fallacy* is that ancient and more recently departed philosophers should be important and even leading members of our inquiring conversations about social life. An act of intellectual courage is needed to rid us of Plato and Hegel. Once they are eliminated, an entire pantheon of outmoded and outdated thinkers, from Aristotle to Kant, will disappear from our radar. This move might also go a long way to eliminating the worshipful attitude intellectuals often adopt to the more productive and visible members of their discourse communities. The caveat here is that some ancient and some modern thinkers (departed ones, as well as some who are still with us) who can be claimed for philosophy are still extremely valuable for us. Marx, Nietzsche, and Wittgenstein come immediately to mind. The corollary intellectual's fallacy is that philosophers as philosophers (and psychologists as psychologists) have anything at all to tell us anymore about the social world. In the wake of the work of sociologists from Emile Durkheim (1995/1912) to Mary Douglas (1988), all the central problems of traditional and contemporary philosophy resolve into (not "reduce to") problems in sociology and anthropology.

Part II: What Role for Philosophy and Biology in the Sociology of Mind and Brain?

Extended Perceiving

The idea that sensing and perceiving are necessary for us to develop an understanding of the world is not new. Jakob von Uexküll (2010/1934) is credited with developing the *Umwelt* concept to provide a generalized model describing an animal's relation to its environment. Simplified to having two branches—a "receptor" and an "effector"—the *Umwelt* emphasizes that these are the key ways in which an entity and its environment are interconnected. Additionally, the *Umwelt* model depends on niche-dependent sensing, whereby different types of organisms may perceive different aspects of the same object in the environment. For example, humans cannot see ultraviolet light while honeybees can. The same flowers in a meadow will look very different to an observing human and observing bee because there are pigment patterns that only are visible in ultraviolet light (that is, they reflect ultraviolet light waves). Finally, this concept reminds us to be careful to avoid homogenizing "the environment" into a monolithic whole when in actuality there are myriad sites for potential interaction, embodied in each organism and even in biologically-relevant objects (like a pool of water).

Ludwik Fleck identified the experience of perceiving things to be essential to the formation of any ideas or thoughts with perpetuity. The ability to perceive depends on training and experience. This also means that we become increasingly less attentive to see things that contradict or are outside "the form" (*Gestaltsehen*). Thought styles are rooted in our capacity for directed perception: "Visual perception of form therefore becomes a definite function of thought style" (Fleck 1979/1935). Additionally, the *way* in which one perceives is of utmost significance; there must be some type of filter applied, what Bowker and Star (1999) would call a "standard" that would promote "classification" of observed phenomena.

The environment—and objects contained in it—being perceived is also a key part of cognition and thinking. Barrett, et al. (2008) discuss the importance of a surrounding environment in providing "cognitive resources" to an entity within it. That environment includes not just other persons but objects that present humans and other animals with "possibilities for action" and "affordances" (Gibson 1979). Understanding affordances in relation to actions can help us understand behaviors in their spatial and temporal contexts. These "cognitive resources" can be used in "epistemic acts," described by Kirsh (1996) as acts used to re-order or arrange information so that it can be more easily processed. These are contrasted with "pragmatic acts," which actually move an agent closer to solving a problem.

Andy Clark (1997) takes this further by explaining that humans constantly gather sensory inputs to orient their bodies in the world as a way to respond to bodily changes (like growing). Using the body and its sensory perceptions, a sense of "where your body is" in space is developed and maintained through constant re-updates from bodily senses; a permanent location in space is too costly in terms of cognitive resources to maintain in static form. This aligns with Karen Barad's theories on iterative, performative materialization (1998); it is through many repetitions that a sense of what is real is created. By hijacking this process, one can "trick" oneself into feeling that one's nose is three feet long if a coordinated set of movements is continued long enough. Clark's insights lead us to ponder what happens to our sense of "self" when our senses are extended beyond the native limits of our bodies. For example, if one were to buy the "Extendable Ears" from *Harry Potter and the Order of the Phoenix* (Rowling 2003) and string them into the next room, conversations there would be able to be heard "as if you were there." Could that idiom actually suggest a shift in reality where the self is in fact extended? Is the hearing part of a person now moved into that other room? Or does the "self" now exist somewhere in-between this room and that?

Discussions of surveillance fit this pattern as well, where devices like video cameras and audio bugs extend power from the executor of authority *into* the room surveilled by allowing them to *be in the room without being there*. This extension of awareness can also be applied to intra-actions that are not specifically sensory, but in a broader sense, perceptive. It is not only the visual or audio inputs that are being conveyed to the authoritative observer, but also the defiance of authority by the transgressor—whether stealing, trespassing, or simply "looking suspicious"—as a social phenomenon is also transmitted.

Intra-action Through Apparatuses

Informed both by Michel Foucault's use of *apparatus* (1980) to talk about social arrangements for knowledge/power, and Niels Bohr's rejection of *apparatuses* as merely passive observing instruments, Karen Barad (2007) presents technoscientific *apparatuses* that are active entities in knowledge production and the observation of phenomena. They are not merely neutral things for probing nature or determining social structures; nor are they simply performative laboratory technologies or social forces: apparatuses are specific material reconfigurings of the world that do not merely emerge in time but iteratively reconfigure space-time matter as part of the ongoing dynamism of becoming.

Can a human serve (temporarily) as an apparatus? A look at the Occupy Wall Street movement's "People's Mic" shows that this is indeed possible. A "People's Mic" amplifies a speaker's vocal message through a one-to-many repetition: the original speaker says a short phrase, surrounding people repeat the phrase, then those surrounding the first repeaters repeat the phrase again, propagating the message out in multiple stages. Developed as a response to bans by local officials on the use of sound-amplifying technologies like megaphones, the "People's Mic" combines cooperation and use of voices to a similar effect. Although this practice circumvents the wording of bans because people working in tandem are not commonly thought of as a "technology," in the context of science and technology studies, these are sufficiently equivalent in composition and application.

In this activity, people serve the function of apparatuses by making it possible for other observers to hear what is being said, despite their being out of range of the original speaker. Although their bodies are not what we'd consider to be externally designed for a purpose (as a megaphone), through the combination of sensory and communication organs, culture and language, and social coordination, these "embodied apparatuses" transmit information beyond its original reach. Assuming that every person-mic hears correctly and uses correct pronunciation, the literal message will be preserved, but even then, the message is transformed (as always occurs in an apparatus in some way, according to Barad) by being spoken with a different voice, with slightly different intonation, or shifted accent. Thus the resulting output is similar to the original, and functionally useful, but still changed (diffracted) in some way. In fact, if one follows the auditory signal's propagation, one might observe a diffraction pattern pulsing in rhythm to the speakers.

In addition to recognizing the capacity for a human body to function as an apparatus[3] to collect, transform, and transmit information through sensory intra-actions, we can also zoom in and observe that each level in our model can serve as an apparatus to the levels above and below it because it mediates sensory inputs through symbolic, sensory, chemical, or electrical means. William James, quoted by Connolly, comes to this conceptualization independently:

The highest and most elaborated mental products are filtered from the data chosen by the faculty next beneath, out of the mass offered by the faculty below that, which mass in turn was sifted from a still larger amount of yet simpler material, and so on (2002, 66).

Therefore, just as "beauty is in the eye of the beholder," so too is a worldview in the (sensory organ) of the perceiver. If we identify the sensory arm of the *Umwelt* as an apparatus, we see that it both shapes and reflects an entity's concerns in the world; things are detected that are relevant to its interests and things that are not, are not observed. Likewise, creating a new ability to detect a phenomenon may create new concerns. Thus are entities entangled with their environments through their entire bodies (which comprise an overlapping, multidimensional *Umwelt* for each type of sensory apparatus) and transforming and being transformed by them with each intra-action.

Mindscape

From this, we can derive a term, *mindscape*,[4] to describe the field of *potential sensory inputs* that are *readily accessible*. The degree to which a particular sensory

---

3   It is important to recognize that abstracting a person's activities to the functions of an apparatus is not undertaken to dehumanize or depersonalize, but to recognize that in coordinated social activities, we engage in complex and transformative transfers of information that shape our world.

4   Eviatar Zerubavel introduced the term "mindscape" in his *Social Mindscapes: An Invitation to Cognitive Sociology* (Harvard 1997). He used the term "mindscapes" to refer

input would be considered "accessible" would vary depending on the context. When asked whether they know the time, most people would have ready access to a timekeeping device (a watch or a cellphone, usually), and most would consider the amount of time and effort needed to access that device to be negligible. If the question were more detailed, such as what the weather forecast is for tomorrow, a website on the internet may be consulted. But the way that such information is accessed could mean the difference between deeming it as "readily accessible" (a Google search on a smartphone) and "not readily accessible" (a hike to a public library to check a computer there, or a wait until the weather report plays on the evening news).[5] In this, as well as in other ways, other entities in a shared or overlapping mindscape are also active in the activity of perception, whether by assisting with obtaining inputs, or in influencing the choices made in how to obtain such inputs.

From this definition of "mindscape," we can derive "mind" as "the process of focusing on a part of the mindscape." By defining "mind" as a process, we gain several heuristic advantages. First, this recognizes the dynamic and intra-active nature of all agential and non-agential participants engaged in perception. Second, this highlights the performative and iterative nature of embodied situating in an environment. Third, discussions of process can be evaluated using alternative ethical systems like virtue ethics rather than being restricted to the tired dichotomy between deontology and consequentialism. The potential for applying new forms of virtue ethics to examine socially constructed material and social technologies holds intellectual and ethical promise.

Expanding to the Social Level

This characterization of cognition as being formed by actions and intra-actions is supported by Vygotsky, who argued the "idea that cognition initially begins by being social and visible and is only later internalized and invisible" (v. Barrett et al. 2008). Just as one can internally differentiate subsystems within systems, creating a new, self-encompassed "system-within-environment" (Luhmann 1995/1984), so too can one expand out from a designated system to embrace part of the environment into a new "super-system." This expandability is key to the model, where after looking at subsystems of a nervous system within the larger system of an individual body, one can expand the scope of analysis to the interconnections between multiple bodies/people delineated as a social supersystem. This represents both complexity and emergence, as defined by

---

to what "we" cognitively share in common. We all (as humans) share some "cognitivities" (Stingl's term), but we also share particular cognitivities that reflect our particular cultures. Zerubavel is keen to avoid equating the "impersonal" shared mindscape with a universal mindscape.

 5 Could knowing how to gain access to an access point for knowledge count as a "meta-mindscape"?

Stengers (1997), with the former being a "conceptual genesis" (our model) and the latter being a "physical genesis" (iterative materialisation of reality).

Clark (2010) identifies a subset of neurons that are involved with "error checking," nicknamed "surprise neurons." These neurons compare new inputs to expectations (developed from past experiences) and show greater activity when unexpected intra-actions occur. It is plausible to suggest that excessive activity by these "surprise neurons" in response to large amounts of unexpected or contradictory data could use up more energy resources and so be a generally unpleasant experience. Therefore, there may be an interest in reducing the energy expenditures of these sub-cognitive routines by making one's environment more predictable.

Several options would be available, all on different levels. For example, if it started to rain, one could use any of the following responses to reduce the "surprise" (and discomfort) of sudden wetness:

1. *Behavioral*: find shelter
2. *Technological*: build a shelter
3. *Social*: change one's outdoor plans
4. *Cultural*: embrace wetness as something pleasant

Likewise, the social environment can produce perceptual inputs that may be surprising and/or unpleasant. To control these, we rely on social norms to select and reinforce desirable behaviors and to reduce and discourage undesirable behavior. For example, it would be very taxing and unproductive to spend time around people who may randomly punch you. Much more pleasant and socially beneficial would be a situation where people did not randomly punch you, either by never punching you or by only punching you for deliberate reasons; social norms of good manners or, in the latter case, a shared culture, would reduce the randomness and surprise. Good surprise, like winning the lottery, also would increase cognitive burdens, especially if trying to decide upon a fair way to split the prize among several people. Again, social norms are useful because they help reduce calculative burdens on actors, especially with regard to sharing obligations, risk, or benefits. They facilitate coordination and energize expectations. We don't have to spend a lot of time—or any time at all—working out a calculus of reciprocity: "In regulated interaction no one has to try to calculate the fair return of cooperative investment in a joint product, taking into account temporal discounts, differential contributions to success and the like" (Sterelny 2008: 380).

Additionally, social norms reduce the chance of unwanted or grossly unexpected encounters in part by making the requirements of reciprocity explicit and by regulating our "preference functions." As a result, the social environment is moderated and made less uncertain by the generation and application of social norms, leading to more constructive intra-actions that promote further inter-connection.

Conclusion

When applied back to the proposed model by Restivo, we see that each level in the heuristic hierarchy of cognition contains entities (whether neurons or individuals) that intra-act with peer entities horizontally and serve as transforming/selecting/ filtering apparatuses for entities in vertically related layers. By building this model on universalizable embodied intra-active concepts, it is possible to expand the model to complex social interactions while also being able to recognize micro-intra-actions down to the cellular level. This approach is intended to challenge disciplinary boundaries and re-frame problems in ways that make it impossible to carve off areas for one set of experts or another; we are defying the "agential cut."

Potential applications of this model extend beyond human social cognition to larger scopes of social institutions as well as to intra-actions sited in and crossing barriers between other life-forms. New subfields like multispecies ethnography offer novel opportunities to develop cross-species modes of communication and intra-action that can produce new insights into ecological thinking (Kirksey and Helmreich 2010). Animal-centered studies, such as on fixed-action patterns (Gould and Gould 2007), could also be pursued in greater levels of detail and in ways that could hold possible connections to obsessive-compulsive disorders (OCD) in humans when a common model of mind and embodied perception is applied.

Discussions of ethics could also be developed in more applicable ways by incorporating a "cultural layering of affect into the materiality of thought" and by recognizing that an ethical sensibility is not a singular concept, but "a constellation of thought-imbued intensities and feelings" (Connolly 2002). For example, the layered, embedded, embodied model of social mind/scape would be well-suited for application to issues like cognitive enhancement because it recognizes that the "enhancement" could occur at many possible juncture points in different ways, promoting more incisive discussion about relevant similarities or differences between types of interventions. In what ways would a chemical intervention be similar to extended behavioural training and repetition, and in what ways would it differ? Would these differences be sufficient to warrant differential legal and/or social treatment? Or would this call into question distinctions made between these interventions that both "enhance" cognition?

Part III: On the Interdisciplinary Imperative for a Theory of Social Ecologies

In Parts I and II we have pieced together some of the ingredients for an interdisciplinary solution to perennial mind/brain/body problems, paradoxes, and enigmas. We have provided a foundation for the model we sketch at the end of this chapter. In this section, we add the final ingredient needed to complete the model—the grounding context for social life, the life of the mind, and our relationships with the world around us.

The interdisciplinary imperative comes, quite simply, from the failure in the course of the twentieth century of our reigning and routinized rationalities (for short, Western rationality). Playing off what Mary Douglas described as the Kantian principle "that thought can only advance by freeing itself from the shackles of its own subjective conditions," we stipulate here that the cultural prisoners of rationalities that have inevitably led to an ecological cul-de-sac can only escape by loosening the canons of rigor.

The end of Western rationality was widely acknowledged during the course of the last century. Milton Yinger, in his 1977 presidential address to the American Sociological Association, said that we were experiencing a planetary-wide civilizational transformation. He pointed out that the countercultures of that period were trying to escape the cul-de-sac through mystical insights, or what we would identify as a variety of "new age" strategies. These strategies represent the typical way cultures deal with the routinization of rationalities. Whereas in earlier periods we witnessed Taborites contending with Prague University masters or English sectarians denouncing the clergy's monopoly on truth, in the last half of the twentieth century science itself was identified as the dying rationality of a system the West had made to travel to the farthest reaches of the planet. One must be careful here not to identify science as the basic reasoning tool of our species with either modern science as a social institution or with rationality as the reigning and routinized system of categories, classification, and rules of reasoning we tend to take for granted in the West as "universal."

Another sociologist, Edward Tiryakian, described "the occult revival" in the middle of the twentieth century as part of a new international "cultural matrix." In Restivo's (1983) earlier study of this topic, he asked if we were experiencing another of those historical episodes in reaction to the end of a rationality, a "sleep of the senses." If the Hermetic philosophers of the Renaissance were in some ways harbingers of the scientific revolution, were there early signs in the twentieth century of a new rationality emerging from the debris of wars, holocausts, atomic bombs, economic booms and busts, and ecological degradations? Perhaps the earliest signs were the physicists' "capitulations to Spenglerism" (the phrase is Smithsonian historian Paul Forman's) in the 1920s and 1930s and the mystical scientists of the 1960s and 1970s. Numerous observers from the 1960s on commented on the closure of science, the evolution of science into a church, and in general on the trap of the reigning rationality (Restivo 1983: 107–8). Restivo came to the conclusion that the way out of the contemporary rationality rut was a secular, no-warranty evolutionary, emancipatory epistemology. The objective was to figure out if it was possible to sustain or raise the adaptive/transformational potential of our forms of life while old ways of knowing and doing were collapsing.

The liminality of our era reaches to most of the fundamental categories and classifications that have guided human cultures for millennia in some cases and for the last few hundred years in the case of industrial societies. This liminality is driving some of the most significant and influential intellectual movements of our era. Nature–society, human–machine, male–female, person–fetus, and life–death are among the powerful dualisms that have become dramatically problematic. The very idea of science (along with those "good" terms rationality, truth, and objectivity) has been embraced by this liminality that threatens to engulf all of our values, goals, and gods. Traditional dichotomies have given way to complexities, non-linearities, and chaotic, fractal, and multi-logical ways of thinking, speaking, and seeing. We have encountered new phenomena across time and space on and off the planet; engaged new ideas, experiences, and values from east to west and north to south (politically, economically, and culturally); and we have endured enormous leaps in our knowledge about how the world around us works.

Our liminal era is producing hybrid ideas and concepts and monstrous entities on a new scale. One day we are accosted by cyborgs, the next day by robosapiens; cloned sheep march with "natural" (cum domesticated) cows and horses; mice are patented; some women sell their eggs, some men donate their sperm. The efforts of the leading spokespersons for this liminality—from Mary Daly as early as the 1970s to Donna Haraway, Gloria Andalzua, Leigh Star, and Bruno Latour—have been exploring new ways of reworking our systems of categories and classifications and at once documenting the changes in worldview our emerging human ecologies are calling forth. Such efforts, now as in all liminal eras, necessarily strike us as awkward, counterintuitive, and obscure to different degrees. In a world of hybrids, monsters, and uncertainties it should not surprise us that the theorists of liminality have produced theories and concepts that are themselves hybrids, monsters, and embodiments of uncertainty.

As we approached the second millennium, the flux of categories and classifications and the proliferation of hybrids and monsters, increasingly came to dominate our everyday lives and the horizons of humanity. These are times that require great courage and imagination to engage, so it is it not surprising that only a few thinkers rise to the occasion. We find interesting and unlikely convergences in this arena of innovative thinking. Bruno Latour, on the one hand, turning from science to metaphysics while David Bohm viewed contemporary liminal dynamics through the very lenses that Latour is seeking to change. Bohm (1976) even championed a verb-based language as one way of coordinating language and reality.

Our discourse exists in a discursive space that has been opening up for a new age of thinkers who have been leading the revolt against the disciplining of the disciplines. We are increasingly faced with problems of survival on a planetary scale that cannot even be formulated in the terms of our reigning and routinized rationalities. In the twentieth century, in the most advanced technoscientific nations, we reached the limits of our categories and classifications. And whatever we could borrow from other contemporary and historical cultures and civilizations has often proven to be too embedded in localities to be useful. At the same time, the encounter with the Others in their multicultural localities has been a catalyst for emerging hybrid categories and classifications that are—or must be if we are to survive—the ingredients of new worldviews, new rationalities, and new modes of knowing, thinking, and action. From a physical perspective, we could say that our narrative unfolds in a Bohm space of infinities of things in becoming. A translation into a commensurable Hegel space (recalling that Bohm was influenced by Marxist and Leninist dialectics) gives us a narrative that is defined not by an aim but by a carrying out of an aim. The result (the conclusion, the resolution) in this sense is not the narrative as a whole but rather the result integrated with the process that brings it to fruition.

We have to bring some old tools to bear on our civilizational crisis. A new way of thinking about brain, mind, body, and culture must be contextualized in a way that links it into the dynamics of the evolutionary stage. The social ecology model and the Umwelt can provide the necessary contexts and engage the dynamics of a robust new epistemological and ontological dialectic.

The *Social Ecology Model*, also called *Social Ecological Perspective*, is a framework to examine the multiple effects and interrelatedness of social elements in an environment. SEM can provide a theoretical framework to analyze various contexts in multiple types of research and in conflict communication (Oetzel, Ting-Toomey and Rinderle 2006). Social ecology is the study of people in an environment and the influences on one another (Hawley 1950). This model allows for the integration (Oetzel, Ting-Toomey and Rinderle 2006) of multiple levels and contexts to establish the *big picture* in conflict communication, health or physical activity contexts. Research that focuses on any one level underestimates the effects of other contexts (Klein et al. 1999; Rousseau and House 1994, Stokols 1996). SEM is primarily a qualitative research model for guiding field observations; however, it has and can also be utilized in experimental settings ... There are several adaptations of SEM; however, the initial and most utilized version is Urie Bronfenbrenner's (1977, 1979) *Ecological Systems Theory* which divides factors into four levels: macro-, exo-, meso-, and micro-, which describe influences as intercultural, community, organizational, and interpersonal or individual. Traditionally many research theorists have considered only a dichotomy of perspectives, either micro (individual behavior) or macro (media or cultural influences). Bronfenbrenner's perspective (1979) was founded on the person, the environment, and the continuous interaction of the two. This interaction constantly evolved and developed both components. However, Bronfenbrenner realized it was not only the environment directly affecting the person, but that there were layers in between, which all had resulting impacts on the next level. His research began with the primary purpose of understanding human development and behavior. Bronfenbrenner's work was an extension from Kurt Lewin's (1935) classic equation showing that behavior is a function of the person and the environment.

Bronfenbrenner (1979) considered the individual, organization, community, and culture to be nested factors, like Russian Matryoshka dolls. Each echelon operates fully within the next larger sphere. Although Bronfenbrenner first coined the phrase *Ecological Systems Theory*, Amos H. Hawley (1950) conducted a significant amount of research in this field as well, along with many philosophers, including his colleague, R.D. McKenzie. Hawley's work on the "interrelatedness of life" in his book, *Human Ecology* (1950), was grounded in Charles Darwin's writings on the "web of life."

SEM is essentially a *Systems Theory* approach to understanding development that occurs in various spheres due to actions in different systems. There are many effects that occur from cross-level influences and relationships between and among levels that SEM addresses. Relationships include parallels or isomorphisms and discontinuities or cross-level effects (Klein et al. 1999; Rousseau and House 1994).

Spheres of Influence

Microsystems consist of individual or interpersonal features and those aspects of groups that comprise the social identity (Gregson 2001) which may include roles that a person plays (i.e. mother, father, sister, brother, child, etc.) or characteristics they have in common. These qualities and factors can be learned in local group settings but many are societal in nature (e.g., ethnicity, gender). In the interpersonal sphere, there are also many components of the individual, including psychological and cognitive factors, like personality, knowledge, and beliefs (Gregson 2001). The individual in his or her own microsystem is constantly shaped, not only by the environment, but by any encounter or other individual they come in contact with. There are multiple, simultaneous influences in child behavior and learning including culture, school, teacher, parental support and education level, involvement in extracurricular activities, etc. Examples of microsystems outside the self also include groups of friends, family, unorganized athletics, and social clubs.

Mesosystems are the organizational or institutional factors that shape or structure the environment within which the individual and interpersonal relations occur (Gregson 2001). These aspects can be rules, policies, and acceptable business etiquette within a more formal organization. Organizations can foster entirely different atmospheres. The organizational component is especially influential with younger, more impressionable employees, as it helps to shape the ethics and expectations of a typical organization for these individuals. Examples include schools, companies, churches, and sports teams. Mesosystems are essentially the norm-forming component of a group or organization, and the individual is an active participant in this group or organization. Bronfenbrenner (1979) also claimed that the richer the medium for communication in this system, the more influential it is on the microsystem.

Exosystems refer to the community level influence, including fairly established norms, standards, and social networks (Gregson 2001). There will likely be many organizations and interpersonal relationships that compose the community, and this web of organizations and relationships creates the community. The community is larger than the meso-; however, it is considerably smaller than the respective nation or culture. The community level in a geographic sense, for example, may be Midwestern or Iowan, while the next level up (macro) would be "American." However, it does not have to be associated with any physical or spatial relationships. Another example could be membership in special interest groups or political affiliations. Exosystems are essentially any setting that affects the individual, although the individual is not required to be an active participant (Bronfenbrenner 1979).

Macrosystems are the cultural contexts (Bronfenbrenner 1979), not solely geographically or physically, but emotionally and ideologically. These influences are more easily seen than the other factors, mainly due to the magnitude of the impact. Examples of significant intercultural effects include Communism, Western culture, Military, Islam, and Christianity. For instance, the macrosystem of Communism is a philosophy based on the belief that wealth should be shared in the macrosystem. A Communist country (in the labeling sense, not necessarily in the Marxist sense), such as Cuba (exo), governs and regulates the environment within which corporations (meso) and society or individuals (micro) exist. Media plays a significant role on all levels, as it communicates information and assists in the development of expectations for all individuals in the respective culture.

Isomorphisms and Discontinuities

Isomorphisms are parallels in the impact on one level and the resulting impact on another level (Oetzel, Ting-Toomey and Rinderle 2006). Researchers studying isomorphic models expect to see an equal effect in both magnitude and direction when at least one influence level shifts. *Discontinuities* are essentially the antonym of isomorphisms. A discontinuity is an effect on one level or group, which produces an unequal impact on at least one more level.

The consideration of top-down effects (McLeroy et al. 1988, Stokols 1996) establishes that environmental effects shape individual behavior. The nested factors are essentially influenced by the external influences that embody these factors. Community and organizational factors often determine how individuals will respond in crisis situations. There is a program called OK-FIRST, which is an outreach project of the Oklahoma Climatological Survey and the Oklahoma Mesonet to educate the community and public officials to help individuals respond in the appropriate way during a weather-related risk. Ethnicity and historical relationships also shape individual conflict behavior. This is obviously true in many situations observed in the conflict in the Middle East. The media additionally plays a significant role in reinforcing these stereotypes. Top-down effects are essentially the most prominent of any of the social ecological components (Oetzel et al. 2006).

Bottom-up effects describe how individuals or community affect higher levels, for example, when individuals form alliances or coalitions to accomplish personal goals. There is also an impact in cultures due to global corporations' presence in some countries. For instance, Google China (even given governmental interventions) has increased the accessibility of information for a wider audience than would be politically and culturally possible. Guerrero and La Valley (2006) recognize emotions are caused by feelings (i.e. anger, guilt, jealousy, greed, etc.) and that these feelings impact events likely to occur. The psychological instability of the shooter in the Virginia Tech incident demonstrates microcosms affecting macrocosms.

Interactive Effects

Interactive effects are interdependent and occur simultaneously at multiple levels (Rousseau and House 1994). For instance in culturally diverse workgroups, there would likely be conflicts between group members, interaction effects in completing the goal of workgroup for the organization, and some learning at the individual level. What role does technology play in cultures, organizations, community, and interpersonal conflicts? McLeroy et al. (1988: 354) noted that the "ecological perspective implies reciprocal causation between the individual and the environment" which essentially defines interactive effects.

To fully encompass the world-context in which our social systems unfold our social ecologies must embrace the *Umwelt*. We introduced the concept of the *Umwelt* in Part II and we rehearse the key elements of that discussion here. The *Umwelt* provides a general model of an animal's relation to its environment. In the simplest model, we have a "receptor" and an "effector," the key ways in which an entity and its environment are interconnected. Additionally, the *Umwelt* model depends on niche-dependent sensing, whereby different types of organisms may perceive different aspects of the same object in the environment. The sensory arm of the *Umwelt* can be understood as an apparatus that shapes and reflects an entity's concerns in the world; things are detected that are relevant to its interests and things that are not, are not observed. Likewise, creating a new ability to detect a phenomenon may create new concerns. Thus are entities entangled with their environments through their entire bodies (which comprise an overlapping, multidimensional *Umwelt* for each type of sensory apparatus) and transforming and being transformed by them with each intra-action. We are led to a three-part model that solves as a first approximation the variety of mind/brain, mind/body, and brain dilemmas and paradoxes that have plagued thinkers for millennia. The model (Figure 1) schematically reflects and explicates Geertz's (2001: 203–17) concept of "culture/mind/brain-brain/mind/culture." It includes the new unit of socialization, a system that encompasses neurons at one end, and interaction ritual chains at the other. This unit is embedded in a social ecology and *Umwelt*. What we can do to formalize (quantitatively and qualitatively such a model is a project we leave for ourselves and others in the future should this approach prove to be viable in solving the mind/brain/body problems. One way to integrate the model would be to reduce each element or ingredient to an information system and link all the systems and sub-systems by way of a circulation of information. We leave the concrete problem of achieving this goal for future research.

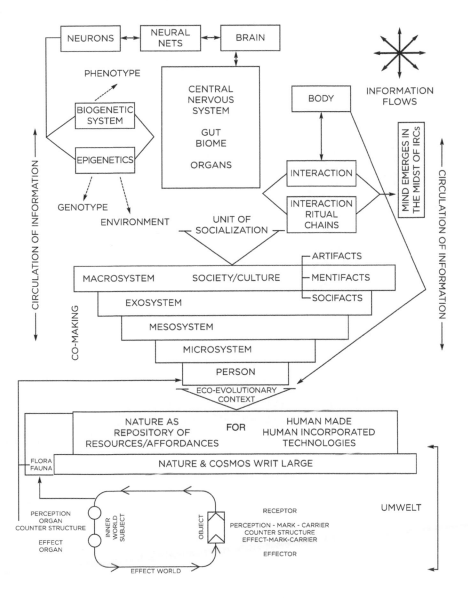

Figure 1   Restivo–Weiss Body/Brain/Culture Model

Chapter 3

Veiling the Modest Cyborg: More Than a New Witness

Sabrina M. Weiss

Reflection

Inspiration springs from unexpected pursuits. What Sabrina Weiss started as a personal narrative about passing as a generic (male) player in online interactions grew into an exploration of Donna Haraway's critique of the "modest witness" ideal in the realm of the sciences. The original questions of "why" and "how" one would project an online persona different from their offline identity mutated into an answer to Haraway's call for a "mutated modest witness" (Haraway 1997: 268) who could transcend the boundaries of politics and technoscience—a "modest cyborg." This being has two components that when brought together derive a meaning greater than their sum—the idea of modesty and the identity of a cyborg. Ironically, the primary inspiration for the cyborg side of this entity was Haraway's own metaphorical being proposed to challenge boundaries and categories. This entity, a hybrid of a hybrid, so to speak, may have the potential to surpass the limitations of the "modest witness" by changing not only the relationship of people to science but of science to people.

Introducing the Modest Witness

The idea of the "modest witness" as a concept in science studies began its illustrious career in the narrative strung by science historians Steven Shapin and Simon Schaffer (1985) in a study that has achieved seminal status in its own field and legendary status in the fuzzy zones of interdisciplinary discourse, *Leviathan and the Air-Pump*: "A man whose narratives could be credited as mirrors of reality was a modest man: his reports ought to make that modesty visible."

This concept, of the scientist (usually a man) who serves merely to report the happenings of reality without injecting himself or his circumstances into the account, has served as a guiding principle for scientific observation. However, critiques such as those from the field of STS highlight the inattention paid to social context, not just of the institutional scientific activity, but of the personal social context of the "modest witness" *himself*. The "modest witness" is a figuration that was born into and is carried by the very matrix that is identified as the Western colonial matrix of power by scholars such as Sandra Harding, Walter Mignolo, and Anibal Quijano. This matrix describes the control of rationality, gender, sex, authority, knowledge, subjectivity, and economy by the creation of distinctions from the linguistic logos of "you" and "I" into the speaking authority (white, male, Westerner) and the silenced Other. Along the lines of this critique, Donna Haraway notes some of the limitations of idealizing the Western concept of the scientific observer as a hermit-like objective watcher, thereby deconstructing the figured world of the Western "modest witness:"

> Witnessing is seeing; attesting; standing publicly accountable for, and psychically vulnerable to, one's visions and representations. Witnessing is a collective, limited practice that depends on the constructed and never finished credibility of those who do it, all of whom are mortal, fallible, and fraught with the consequences of unconscious and disowned desires and fears. (Haraway 1997: 267)

To answer these needs, Haraway calls for a "mutated modest witness" who is able to see her/his surrounding context and who "diffracts" information (thus actively transforming it) rather than simply reflecting what is observed. "It is a kind of modest witness that insists on its situatedness," this evolved being, and it realizes that it builds the world around it even as it explores and reports on it, not as an "it," but as a person beyond "she or he." The inhabitants of this "figured world," in other words, are aware of its figuratedness, and enabled to play with figures and figurations without silencing one for the other. As a consequence, the type of being that can navigate the ever-shifting landscape of a changing reality must itself defy boundaries, thus the call for a cyborg to take up this mantle. So first we shall find our cyborg, then we shall teach, or be taught, modesty.

Becoming a Cyborg

In its essence, the idea of a cyborg confounds classification even before it exists; this is the core of Haraway's definition of a cyborg: "A cyborg is a cybernetic organism, a hybrid of machine and organism, a creature of social reality as well as a creature of fiction" (Haraway 1991: 149). As an *autopoietic cyborgological law* (an ironic twist of Haeckel's biogenetic as well as Elias' sociogenetic laws), we can conclude that the ontogeny of a cyborg recapitulates its self-perpetuating socially constructed phylogeny: which came first, the boundaries or the desire to transgress those boundaries? Although typical images of cyborgs involve organisms (usually humans) bearing machine-like parts to augment or enhance themselves, it is important to remember that technology need not be made from metal; indeed, one should appreciate that social institutions and traditions are every bit as technological as are objects of metal, and no less unyielding to resistance. However, in using the anachronistic "alchemical concept" of transmutation along with the telematic notion of information and computer/communication technologies (ICTs), we want to start with our example of "cyborg transmutation" as being centered on a human-computer interaction because it is more apparent to the observer. This will empower us to later identify the social technologies constraining and enabling us in our activities, science-oriented and otherwise.

An Auto-Ethnographic Sketch of a Transmutation Narrative

As a child, Sabrina Weiss grew up with computers at home, and, under the influence of her father, she enjoyed playing video games. From a young age, she says,

> I also became aware of the social norms restricting us to activities deemed appropriate for our random lot in life: my preschool class had insisted that I could not become a doctor because I was 'just a girl,' and therefore had to settle for being a nurse. I also was teased for bringing 'strange Korean food' to school in my *Thundercats* lunchbox (which included *myulchi*, soy sauce-sautéed dried whole anchovies). Although my parents supported my aspirations and encouraged me not to be bothered by these artificial limitations, I advanced through life keenly aware of stark divisions between people that were reinforced as impenetrable but which I constantly stepped over, around, or even through without realizing. This acceptance of the fluidity of identity may have been a factor that allowed me to pursue typically male, intellectual activities: studies in science and math, science-fiction and fantasy reading, tabletop role-playing games, and video games. I was used to playing male protagonists in computer games because prior to Lara Croft, there were few female leads; it rarely bothered me, and in fact was liberating in a sense to recreationally adopt other identities.

Her first foray into online interactions in a gaming environment happened during her last year of college while playing *Diablo* (Blizzard 1996) online through Battle.net. When her adventuring partner found out that she was female (because he asked and she "naïvely told him the truth"), he proceeded to engage in what would count in the offline world as heteronormative masculine "chivalrous" acts like "opening catacomb doors as we hunted zombies and demons and saying, 'after you.'" He also gave as a gift an item he earlier had refused to trade for (before he knew that his online partner was a girl), another commonly gendered behavior. Although this was a much milder experience than many female gamers have suffered, it made evidently clear the impact that gender projections online had. In later online interactions and virtual ethnographic research, mostly on text-based social blogging sites like LiveJournal, Weiss experimented with adjusting her typed "voice" to reflect at least a neutral gender leaning, if not a slightly (geeky) masculine flavor. The result of this was that she was rarely even asked about her offline gender, with most people online referring to her by default as "he." From the point of view of subjectivities that one can experience online, she notes that this

> gave me a welcome position of unremarkability (not invisibility, as I was an active participant) in the realm of online flirting and soliciting, made more apparent when one poster 'erroneously' suspected I was female and promptly tried to hit on me; I laughed it off brusquely (without actually lying), and he backpedaled furiously, apologizing for his 'mistake.'

Years later, she found herself coaxed into trying *World of Warcraft* (Blizzard Entertainment 2004) with a friend. Drawing on her previous Battle.net experience and growing academic and research background, she carefully considered options at character creation since now there was a broad range of choices for avatars, including gender, race, class, and appearance. Based on various observations and researched literatures about geek and Internet culture, she had developed a rough hypothesis for gender projection online. Language used in typed communication, she found, was a strong factor in transmitting gender, but more powerful and insidious were the traits of an avatar because they conveyed a great deal of passive and indirect information insinuating at the person behind the screen based on social expectations of males and females. From the point of view of a gamer's subjectivity she was interested not in disproving the accuracy of the social expectations, but in using them to her advantage to avoid the ethically distasteful trap of being asked directly about offline gender and subsequently forced to choose between lying to protect herself or risking exposure to discrimination or harassment by telling the truth. In the end, through avatar selection—a long-tusked Troll with a mohawk and an enormous humanoid bull—careful development of online "voice," and enjoyment of a more tolerant and progressive social climate, she managed to successfully "pass" as a male player in the year that she played *WoW* with no one feeling a need to ask questions about identity. Because it is relatively easy to accomplish this feat of "genderbending" online, it can be relatively difficult to get solid data on the phenomenon. This is because successful genderbenders who are invested in their identities would often choose to stay *incognito*, remaining nigh-invisible to surveyors and surveys. It is therefore problematic to obtain robust and reliable data on the salience of this type of gamer subjectivity, but it seems reasonable to conjecture that it is not an isolated nor minimally significant phenomenon simply because there is not much data on it.

Analysis

To richly engage with this topic, we shall examine this narrative from several angles: First, text as the primary communication method used online will be examined in how it contributes to a person's image of another. This raises the question of how it contributes to image formation between people. The idea of schemata will be integrated into this discussion to emphasize why there is such a reliance on shorthand stereotypes to classify people. It is, indeed, plausible to argue that nowhere but in early and currently developing online-gamer culture is practice so readily reducible to text-based practice and, subsequently, open to text-reduced methods of analysis (such as deconstruction). It remains to be seen how well future knowledge&technology communities of gamer cultures will either mimic the complex practices and communication techniques of the offline world or, more excitingly, even create non-textual discursive practices of their own. Second, genderbending in massively multiplayer online role-playing games (MMORPGs) will be discussed and contrasted with *"passing."* Third, the use of technology to achieve goals of genderbending or *passing* will be examined in online and face-to-face contexts. Perspectives on the term "modest" will then be discussed, followed by the presentation of the *"modest cyborg."*

The primary active form of communication used online is typed text. Although voice chat is increasingly used in cooperative and competitive games, it is essentially a direct transfer of offline inputs without transformation or diffraction—it is a telephonic technology simply added to an activity. Christian Schmieder (2009) classifies four modes of game-related communication—visual avatar, text, guild websites, and voice chat—into two main categories: organized by the game software (avatar, text) and by gamers (guild, voice). This supports the differentiation of voice chat from text or game graphics. Voice-enhancement software is available and succeeds to varying degrees, but we will not be covering this separate issue here.[1]

Text-based Communications

Awareness of text-based projection of identifying factors is not particularly new, but studies of earlier text-only online interactions, such as multi-user dungeons (MUDs), have emphasized the important contribution that writing styles make to the image of a person. Lori Kendall (1996: 220) notes: "In the limited bandwidth of text, typed conversation is the only means of communicating gender identity, and communicating it in a complex or nuanced way can be very difficult." It is easy to forget how much that is assumed in "normal" face-to-face interactions becomes lost (perhaps intentionally discarded?) when communicating online. In addition to an appearance that is usually gendered and ethnically suggestive, subtle cues of posture, gesture, and facial expression convey information that contributes to an observer's picture of what that person is like. Again, much of this can be covered by learning from the classic theorists of social interaction from Mead and Simmel to Vygotsky, Geertz or Goffman. We do not need to reinvent the wheel here. But

---

1   It is worth noting that from Weiss's personal observations, a majority of voice-enhancement users are male. They use it for two primary purposes: 1) to deepen a pre-pubescent voice to avoid ridicule for being young or young-sounding; or 2) to support an illusion of being female to another male user for the purposes of extracting gifts from him.

we do wish to emphasize that prior to the advent of webcams, a lot of the implicit technologies and forms of social interaction (more often tied to the construction of self and identity politics than not) were seldom available to the average online user. Instead, people relied solely on text, both in substance and in style, to infer something about the person they talked to. This is difficult enough in one-on-one conversation; in a situation with many people all conversing at once, the amount of attention that can be paid to any one person drops significantly. Instead of seeking an in-depth understanding about someone, people end up relying on shortcuts and frameworks to quickly sort people, establishing a particular type of figured world of its own. One model for this refers to a psychological schema, a script[2] by which the world is understood (Widmayer 2012). When confronted with new information, a schema allows it to be quickly processed with a minimal amount of thought—accepted if it fits the schema, and rejected if it doesn't. Once a schema is established, reinforcing information is assimilated invisibly, and so long as the schema is not disrupted (necessitating a new reassignment of a schema), contradictory information is thrown out without much distress. This phenomenon was described in an interview with Shannon McRae (1996):

> When I do something that confirms your impression ... you suck it up without even thinking because ... because schemas are our brain's way of being efficient. You only have to learn farmer once or twice. then you can just call it up. and you can add new stuff to it to expand. ... But ... mostly you call up greenjeans ... So ... our brains are trying to conserve tasks ... can only handle so much at once. So we construct and call upon schemas for a great deal of stuff. one of our schemas is gender [sic].

This is significant for several reasons. First, the deep distrust expressed by many about interactions originating on the Internet can be connected to the intersection of "limited bandwidth" discussed by Kendall and the over-reliance on schema described to McRae. People are uncomfortable when they don't feel like they know what the "other person" is like, and the limitations of the text-based medium strip away all of those subtle physical cues we usually rely on to generate an image of someone. Second, through the process of identifying *why* people are uncomfortable with restricted forms of communication, we can identify how and the extent to which people rely on those other modes of communication.

---

2 There are numerous theories on scripts, schemas and such. We do not want to enter into a debate of all or any of them here, but merely point out two very different lines that this can be explored by other than psychology: On the one hand, the inherent mechanisms can be conceptualized philosophically with theories of reference or holism such as Paul Grice's work on implicature (1989), Robert Brandom's inferentialism (1998) or Phillip Pettit's holism (1996). On the other hand, there are anthropological and sociological works on cultural scripts such as Wierzbicka's on explicatures (2003), on rituals such as Victor Turner (1969) or Randall Collins's interaction ritual chains (2005).

But how do people derive enough information from text to trigger a gendered schema at all? Based on an algorithm developed by Shlomo Argamon, Moshe Koppel, Jonathan Fine, and Anat Rachel Shimoni (2003), a website called "The Gender Genie" claims the ability to assign gender ratings to text.[3] This algorithm focuses on the use of connector words like "with," "around," "many," and the use of passive verbs like "are," "is," and "was." To determine the gender orientation of text, these connector words are explicitly (in the Gender Genie) or implicitly (by most readers) sorted into "feminine" and "masculine" and tabulated; a larger portion of words in one category indicates that gender. Within a generally Western culture-specific context, a masculine voice tends to be more blunt, forceful and decisive while a feminine voice demonstrates more personalization ("I think"), qualification, and description. This type of sorting alone would not be sufficient to result in semi-rigid gender assignments, but when paired with the limited bandwidth problem described by Kendall (1996), there arises a strong reliance on gender caricatures as a way to cope with the lack of supplementary data. This results in the observed tendency for people to judge and be judged in gender identity by seemingly sparse information. Although it is valid to critique the underlying mechanics of the "gender algorithm" because they are embedded in certain Western linguistic norms, it is important to recognize that the interactions described earlier were situated, for the most part, in places observing those norms (with participants serving as the anchors to those locales), resulting in the rise of the corresponding schema.

Avatar Communication

Another level of communication occurs on a visual level for online environments that allow avatars or other visual communication, such as user icons. While not yet explored as an example for a dynamics that Stingl (2012), in his work on medical imaging technologies, has called the *narrative dialectics of technoscientific seeing,* the interplay of knowledge practices and narrative of selves can certainly be cast in this visual frame of reference (and see Stingl forthcoming b, 2012a, with Weiss 2013). Indeed, the use of avatars for self-expression and as social mediators has been studied in recent years on various levels, with special attention paid to the types of passive communication that occur through the choices made in the creation process. By choosing an avatar of a different gender, a player can engage in a type of "genderbending" activity. It is important to note that there is a difference between "genderbending," expressed as a player of one gender simply choosing an avatar of a different gender but not necessarily adopting a different player persona, and "passing," expressed as a player of one gender intentionally promoting a perception of being a player of a different gender through selective interactions. In the former,

---

3   At the point of testing, the Gender Genie assigned Weiss a score of 2753:1843 M:F and declared her to be male.

it is a fairly straightforward data collection process because most online RPGs clearly distinguish between male and female characters (the high level of dimorphism between male and female renderings is a topic covered extensively in discussions about media and video games). However, in the latter process we see a more nuanced exploitation and utilization of schema and other social cues like gendered text language to support the activity.

Schmieder (2009) places passive visual communication through avatars at the end of the spectrum as "most anonymous" and "least communicative freedom." The perception of lack of freedom depends partly on how many options like race and class are available to gamers when designing an avatar. Nick Yee, at the Palo Alto Research Institute (PARC), has conducted extensive research regarding avatar demographics in both *Everquest* (Sony Online Entertainment 1999) and *World of Warcraft*. In earlier studies, Yee (2001) tabulated the avatar gender[4] of both female and male players. He found that in a hypothetical pool of 1000 players, 840 would be male and 160 would be female players (new data show that female players may comprise around 40 percent of the online gamer population today). However, male players were much more likely to have a female avatar than were female players to have a male avatar, so much so that 1 in 2 female avatars would be played by a male player while 1 in 100 male avatars would be played by a female player. The player and avatar gender distributions came to 84/16 and 65/35 male/female, respectively. The phenomenon of male players choosing to play female avatars claimed a fair amount of attention from the researchers. Through surveys, various reasons were identified, including visual appreciation (many MMORPG games are played with a perpetual view of the avatar's rear end), aural appreciation (female grunts and moans of exertion), size of avatar, psychological warfare (implying shame felt by a male player beaten by a female avatar), and exploitation of gender norms (male players often give gifts to people they perceive as female).

By comparison, there has been little, if any, examination of why (a very few) female players choose male avatars. This is perturbing on both a scholarly and a personal level: there often seems to be a disproportionate amount of attention paid to male activities and male motivations over the corresponding female versions, and as a female gamer who chooses male avatars over female avatars, it feels disconcerting to be ignored and deemed insignificant.

It is notable too that even in surveys of player motivations, question and response categories failed to recognize significant and unique responses by female subjects. For example, in a survey of *Everquest* players (Yee 2001), gamers who chose avatars of a different gender were asked why they did so. Their responses were sorted into categories such as "visual appearance," "role-play purposes," "to gain advantage in-game," and "gender exploration." An additional factor was

---

4   We feel that "gender" is more appropriate a term than "sex" for computer avatars that are not engaged in sexual activity and reproduction as any visual aspects are purely performative and socially based, rather than materially/biologically based in reproductive activity since genitalia are rarely present on these avatars.

"the Tomb-Raider effect," defined as "the appeal of an aggressive yet attractive female." Two responses by female players were cited because they "didn't fit neatly into any of the categories mentioned already":

> I thought the blond barbarian quite the hunk. And since it was obvious that none of my girls would get hooked up with one, I decided to play him myself.
> [...]
> I created a male character strictly for the purpose of getting completely away from everything. It is the last place anyone would think to look for me. There are also times when I just don't feel like putting up with some dork who's found out that I'm female in real life following me around hitting on me. (Yee 2001: 41)

It is perplexing that the top response regarding visual attraction ("quite the hunk") was not coded to fit one of the existing categories. Perhaps "visual appearance" was designed to report only a sanitized, purely pixilated aesthetic rather than any sort of prurient interest, which was reserved for the "Tomb-Raider effect." The bottom response calls attention to a lack of awareness in the construction of response categories, as there was no category recognizing a desire to not be sexually solicited, or even more broadly, to simply "get away from reality" ("whose reality?" is a valid question here) for a while. The explicit recognition of a "Tomb-Raider effect" further emphasizes the male-oriented construction of this survey and report since this is an effect predominantly experienced by male gamers as they enjoy a challenged female gender norm in the form of a controllable avatar. Additionally, this allowed the category of "visual appearance" to be reserved for purely graphics-oriented aesthetic evaluations (another angle that could be considered male-oriented because of the coding/processor/video card aspect), leaving female players who enjoyed "hunky" male avatars out of tabulation. A connection could be drawn from this attitude back to the culture of mentally pure, sans-sexual constructions of scientific research (Haraway 1997: 28). Ironically, the female genderbenders in this study are the most effective demonstrators of cyborg-like behavior of anyone involved in the research and highlight the need for Haraway's evolved modest witness through their category-disrupting activities.

Genderbending vs. Passing

With this in mind, it is important to note a difference between what most of the genderbenders in the afore-mentioned study did as opposed to what Weiss did in her online experience. While in both cases participants opted to play avatars of different genders than those attributed to the offline person, there was little discussion in the study about people who actually attempted to *pass* as players of a different gender. Even without actively attempting to *pass*, however, female players in the *Everquest* study who genderbent through male avatars did observe differences in how other players treated them, suggesting that an inadvertent *passing* effect resulted from their selection of a male avatar. In general, they were treated with more respect and were seen as more competent:

> When I play my male characters, other male members of the party will listen to me better, take me more seriously. *In my male form I could give orders and have them listened to, where as a female, my characters aren't always taken quite as seriously.* (Yee 2001: 41; emphasis added)

This shift of perspective brought new awareness when the player returned to a female avatar, whose appearance elicited strictly gendered behavior from other players, such as gift giving. The player interestingly called the male avatar a "body" upon reflection: "I've enjoyed the higher level of 'respect' for my abilities that seems to come with playing in a male body" (Yee 2001: 4).

Conversely, when a female player's gender was revealed through casual conversation, behavior by male players was reported to have changed drastically:

> I was in a group I had worked with for a while. I was playing my male paladin and trying to be the group tank. I was pulling the mobs too (in a rather dangerous zone). After a while, a conversation about our home lives started and I made a comment about my husband. *Immediately the guys in the group asked me if I was really female in RL* [real life]. *When I confirmed it they started sending out another male character to pull the mobs. I found the whole group suddenly expecting me to do less melee.* I'm not sure if they became protective of me or if they just assumed that a female would be less capable in the role of a tank. (Yee 2001: 4, emphasis and clarification added)

What had been a stable social interaction was forced back into conventional offline structures, with males insisting on "protecting" the (virtual) body of a female, questioning her ability to serve a role as the vanguard protector, and letting offline considerations override proven strategic understandings that had been established earlier. Thus this type of transgression is a fragile construction that can be obliterated when the masculinity of the player is called into question.[5]

Meanwhile, some male players who passed as females gained insights into challenges uniquely faced by female players: "I never realized how irritating it can be to have to put up with unwanted advances" (Yee 2001: 4). Although many respondents experienced different treatment when using male or female avatars, they did not necessarily intend to portray the "person behind the toon" as that gender, but were often assumed to match the gender of the avatar unless contrary

---

5   It is not clear whether this is a necessary result or one that merely reflects biases against online interactions. The convention of labeling offline activities as "real life" implies that all activities online are un-real and therefore inconsequential. We have intentionally avoided this bias by using the more parallel terms "online" and "offline" to focus on the more relevant distinction.

evidence was observed: "I learned that I use too many hehe's and :)'s to play a guy" (Yee 2001: 46). There were no cases discussed in Yee's *Everquest* study where players intentionally attempted to *pass*, but there were some cases where players allowed others to make assumptions about their gender identity without dissuading them, figuratively pushing against or even poking small holes in the barriers as proto-cyborgs. Even without intentional transgression of barriers, these players gained insights from their explorations and became more aware of their situatedness in the cultures of both online and offline life.

Compared to the MUDs discussed earlier, there is another layer of abstraction embodied by the avatar that passively communicates information. Depending on how one views the mind-body connection in face-to-face interactions and the difficulty of passing in such, it could be argued either that the MUDs or the avatar-based games more closely mirror offline interactions. It could be that MUDs, while conveying less information than do face-to-face interactions, do not add extra levels of abstraction that can be manipulated by a user/person the way avatar-based games do. On the other hand, it could be argued that by giving players a virtual body to control and ornament, avatar-based games are closer to face-to-face interactions because they mimic our default modes of communication with faces, arms, legs, gestures, etc. This field of resonance and ambiguity is a perfect haven for a cyborg.

Offline Examples of Passing

Genderbending and passing are not unique phenomena restricted to online interactions. In fact, there are numerous examples in literature and throughout history of both males and females challenging gender norms and identifications. Examples here will focus on female-to-male passing to balance out general under-representation and to highlight some common patterns between online and offline female-to-male passing.

The Chinese epic ballad of Hua Mulan told the story of a loyal daughter who went to war in her father's place (Edwards 2008). Dressed in his armor, she was able to pass as a male in the military and over time earned the respect of her peers and superiors for her intelligence and courage. In the end, she declined an offer to be appointed to a high official position and returned home to be with her father. In the poem, her "true" identity was unknown until she revealed it to her military escort by dressing up in her old clothes at home. Also known in the United States as a Disney movie with a stirring soundtrack (*Mulan* 1998)—one song is ironically titled, "I'll Make a Man Out of You"—this story has been used as the basis for several movies in China as its story has resonated across the centuries.

In J.R.R. Tolkien's high fantasy series, *The Lord of the Rings* (1955), female characters for the most part were relegated to idealized and distant depictions of women who were more part of the scenery than of the plot. One exception to this was the character of Eowyn, niece of the king of Rohan. Defying her uncle's order to stay home during a climactic battle, she dressed in armor and rode with the army to face the enemy, passing as a nondescript man the whole time. She found herself face-to-face with the Witch King, an elite general in the enemy's army, who had bragged that "no man may hinder me." Her deception proved to be a vital turning point in the war, as she revealed herself as a woman and destroyed the Witch King, much to his surprise. After recovering from her wounds in battle, Eowyn retired from fighting and married another hero of the war; apparently Middle Earth was not ready for such a woman to continue to fight.

History reflects (or inspires) this trend of women passing as men to seek honor or accomplishment through battle. Some remains from the Civil War have been identified as belonging to women who were dressed in men's clothing (Blanton and Cook 2003). In Albania, the practice of "sworn virgins" allowed families without a son to retain control of property by turning one of their daughters into a son (Bilefsky 2008, Grémaux 1994). Fortunately, this only required a change of attire; clothes really did make the man in Albania. This option was also available to daughters who did not wish to marry any approved suitors; they were allowed to stay single if they socially converted to being men. Although this practice is no longer being perpetuated, the last of these sworn virgins have noted several benefits unique to this practice (as opposed to simply increasing the status of women in their society). In an interview, one of these men commented that "being a woman made her a more compassionate man. 'If the other men were disrespecting a woman, I would tell them to stop'" (Bilefsky 2008). In addition to representing a tangible societal benefit in moderating hostile behavior, the wording of this statement reflects some interesting cyborg-like fluidity and acceptance in a state of being that simultaneously exists in overlapping ideas. By having the option of changing one's situated perspective, whether by adopting the manner, dress, or even "body" of a male, female actors gained both a new perspective on a completely different mode of interacting with others in the world and a new ability to affect the experiences of others by altering established gendered interactions; this combination of altered *perception* and *ability* constitutes a new type of agency. Thus a single act of transgression can produce a cascading effect of changed perspectives and interactions as the very pathways used for reinforcing set patterns become appropriated instead for disrupting those very patterns.

Social Technologies and Modest Witnesses

Where in these stories is the technology and/or techno-scientific practice? Attire, for example, is a vestimentary medium of communication as much as it is a techno-scientific practice that concretizes material semiotics. As such, they appear in multiple locales that connect "socio-corporeities" and "sociomaterialities" in the form of networks (Patrick Carroll 1996: 142):

> ... to designate that through which otherwise separate phenomena, spaces, or
> practices, are connected together. The network subsists in heterogeneous forms:

social, material, discursive, natural, and cultural forms; in nodes and linkages that are both 'human' and 'non-human'; in roads, vehicles, reports, statistics, books, and pamphlets, modes of dress and address, and persons and objects that move in socio-material space, as well as in meta-discourses on rationality, sexuality, morality and economy, which subsist with and through these persons, books, objects etc.

Attire, whether clothing or armor, is a key technology used by these women to pass as men; by changing their appearance, they were able to control their identity as projected to/received by the people around them. "But they aren't *really* changing their gender, are they? They're just covering it up!" argue critics. But it is important to recognize that "gender is relational," according to Haraway,[6] and that gendering is a mutually active process whereby beholder and beholden interact with each other using technologies, social or material, to generate between them a classification. Julia Serano (2007) argues similarly that although there is a tendency to see naturally existing categories of "male" and "female" as preceding any classifying activities, in actuality we "compulsively" and actively distinguish between women and men. Schemata are part of this, connector words are part of this, and appearances moderated through avatars or clothing are part of this. As we saw when discussing MUDs and the stereotypes people would utilize to classify others, "we tend to make the call one way or another no matter how far away a person is or how little evidence we have to go by" and rather than relying on inductive processes to collect observations into a final verdict, we deductively project expectations about maleness and femaleness onto each individual we encounter (Serano 2007).

Perhaps it is an intrinsic part of human, or even organismic, nature to classify compulsively and incessantly, or perhaps it is a social adaptation shared across societies. What matters is that regardless of what one does, one will be sorted into one group or another by surrounding people. Although Serano argues that "the public is the primary active participant by virtue of their incessant need to gender every person they see as either female or male," this perspective pushes a little too far in the other direction. Just because most people are unaware of the mutual gendering activity that occurs in every social interaction, it does not mean that they are not complicit, or even active in this process. As with the Sorting Hat of Hogwarts in the "Harry Potter" series (Rowling 1997), where from an outside perspective it appears that sortees simply sit there neutrally while the hat issues a proclamation, what actually occurs is an active dialogue between the hat and the student:

---

6   Relational understanding in this context means that "understanding of self and other, [...] arises from a confluence of propositional, affective, and tacit forms of knowledge about [... *the problematic issue in question*,] and one's own situatedness within." (Perry/ Shotwell 2007: 34)

'Hmm,' said a small voice in his ear, 'Difficult. Very difficult. Plenty of courage,
I see. Not a bad mind, either. There's talent, oh my goodness, yes—and a nice
thirst to prove yourself, now that's interesting ... So where shall I put you?'
Harry gripped the edges of the stool and thought, 'Not Slytherin, not Slytherin.'
'Not Slytherin, eh?' said the small voice. 'Are you sure? You could be great, you
know, it's all here in your head, and Slytherin will help you on the way to greatness,
no doubt about that—no? Well, if you're sure—better be GRYFFINDOR!'
Harry heard the hat shout the last word to the whole Hall. (Rowling 1997: 90–91)

When Harry Potter was sorted, the hat mused about putting him into Slytherin
because of his immense potential for dark magic, but Harry begged to not be
placed there. The hat responded to his pleas, sorting him into Gryffindor instead.
Similarly, the process of gender projection/gendering is a negotiated series of
proposals and rejections that take place invisibly to most people, as coordinated
as a ballroom dance floor. In this case, "the public," embodied by individual
beholders, is the same "modest witness" found in science—claiming no personal
identity, stake, or bias and "just reporting the facts, Ma'am (or Sir)."

It does not need to be this way. What these examples show is that one can
intentionally and consciously step into the dance with an awareness of the rules
and an ability to play by those rules to achieve one's goals. Mulan and Eowyn
knew this and found glory in battle. Albanian sworn virgins were able to engage in
this way with the explicit cooperation by their society, a rare and fortuitous case.
The genderbenders in MMO games are like the novice dancers on the sidelines,
practicing and experimenting, learning as they go, some realizing that they have
aptitudes they had not previously recognized, some gaining new insight into how
much agency one can have even while following the prescribed moves. This
awareness paired with ability qualifies a person as a cyborg with a new agency.
Now that we have found our cyborg, we shall look for modesty.

On Modesty

In the context of the "modest witness" of science, modesty suggests an invisibility
or a reticence about one's personal circumstance. A complex word, modesty has
many facets, not all of which are seen in this usage. Surely, by placing his name
on a research document, the scientist is not exercising modesty in eschewing pride
in his accomplishments! Other definitions of modesty include: "freedom from
vanity," "regard for decency of behavior, speech, or dress," and "moderation."
Synonyms include: "self-effacing," "humility," and "discretion." Modesty has
also carried moral weight, especially in religious contexts. Thomas Aquinas,
in *Summa Theologica* (2008/ca 1273), wrote about the importance of modesty
in attire, especially with regards to clothing that could provoke lust in others.
However, he made allowances for some cases, usually of practical or charitable
natures. For example, while makeup was generally considered sinful because it
was falsification, hiding a disfigurement was acceptable. And while cross-dressing

was considered sinful when engaged in for sensual reasons, a person wearing clothes of another gender to avoid harm, hide from enemies, or compensate for lack of proper attire would be accepted. The motivations of our examples fit the latter more than the former and would likely have garnered Aquinas' acceptance, if not applause.

In considering women, who are often admonished to exercise modesty in their dress, often to avoid provoking thoughts of lust in men, it could be argued that passing as a man through the use of attire and other cues is the pinnacle of modesty. For as long as a woman is identifiable as a woman, her presence can inspire imagination in the mind of a man (or a woman so inclined) that make all the clothing in the world ineffective. To completely reject objectification through the immodest witnessing gaze of men, a woman must pass as a man, effectively turning the tables back onto the viewer and making him question his own interests in light of frequent taboos against homosexuality. Simply wearing men's attire is not enough, however. Although coveralls are a unisex uniform on naval vessels, U.S. Navy regulations require women to wear their belts in the "woman's" orientation (reversed from men's) and forbids women from completely shaving their heads; "passing" apparently poses a threat to good order and discipline, even in modest attire.

Looking at modesty in the sense of self-effacement or humility, a cyborg who willingly sets aside one aspect of identity to adopt another for the purpose of gaining access to an activity as an unremarkable common member embodies a deep personal sense of modesty. To recognize that one's status in one context does not matter in another represents a level of social maturity we expect from people who cross communities. It would be considered inappropriate to focus on military rank in a civilian classroom and uncouth to trumpet one's golf score in church. Yet when interacting in an online environment where offline bodies do not matter, most people insist on dragging in irrelevant gender classifications. "Everyone is in 'drag' on MUDs; being more or less female has no relationship to one's gender identity off-line," notes Kendall (1996. 219). The cyborg, then, is both modest and wise in accepting this reality and willingly reformatting her/himself to fit into a different context. This outlook is also inclusive and empowering as the agency of surrounding people is embraced and even celebrated much in the way an anthropologist strives to embrace a culture.

Modest Witness Meets Modest Cyborg

We now return to the modest witness. Although this paper has focused on examples of modest cyborgs, the context from which these cyborgs arise is filled with modest witnesses blithely singing along like a Greek chorus: "women cannot be doctors," "you don't talk like a guy," "we don't know how to classify a woman who wants to play a hunky barbarian." By singing the standards in monotone echo, these witnesses reinforce them while denying responsibility for their compliance. The researchers with insufficient categories in the *Everquest* study, the Witch King who defied all men and therefore thought himself safe, that class of preschoolers from long ago, and every member of the public, both offline and online, who insists that gender is apparent and inherent all stand in this choir with hands over their eyes and ears.

But this is not just a problem that confounds scientific communities. The online/offline difference of practices (of gender identities and their representations) is similarly at stake thanks to people participating in online communities, list servers, etc. Anne Fausto-Sterling, developmental biologist and seminal voice in gender studies, has frequently remarked on her experiences in online fora and listservs when addressing the issue of gender. We must ask whether or not the world of gaming and the world of online science communities are equally "veiled." And not only is the issue of gender and online textual practices a point of interest; despite all efforts to guarantee equality and unbiased procedure through (blinded) peer review, research grant applications have been subject to biases in view of race (Baskin 2011, Corbyn 2011). Consequently, Haraway emphasizes the need for *diffraction* and not just *reflection* in witnessing; one who cannot see the light has no chance of transforming it into a spectrum that rejects a dichotomy of light/dark. Yet this is the entity upon which society must rely for matters of fact and truth. The limitation of the modest witness is not in a lack of *modesty*, but in an excess of *witnessing* while denying participation in any other activity.

What the modest cyborg offers is not merely a critique of boundaries, but a redefinition of the activity in which s/he is admonished to be modest, thus evolving the very definition of modesty to be used. A cyborg may witness, but also may question and participate and create change; being unrestricted by one's title has its benefits. In this spectrum of available activities, a different definition of "modest" may be applied to each as appropriate to the situation, recognizing that modesty is not merely covering up, but being willing to strip away unnecessary baggage.

Another boon of the *modest cyborg* is the inducement of social evolution through a version of Haraway's *recursive mimesis*:[7] "The object studied and the method of study mime each other. The analyst and the analysand all do the same thing, and the reader is sucked into the game" (Haraway 1997: 34). The cyborg passes by imitating enough of a desired group to infiltrate as an unremarkable member. But then the limits of that category are challenged as the cyborg accesses experiences and ideas that are foreign to that group. Because the cyborg is accepted as part of the group, other group members in turn mimic the cyborg's slightly altered patterns as part of a general socialization pattern of ritualistic imitation. Conflicting behaviors result either in destructive interference or in disruptive technologies (i.e. Techno-scientific practices), and, in either case, reinforcing behaviors are magnified, which (re-)stabilize the now transmuting

---

7   Under conditions of method meaning a technique of seeing or a *gaze* and a narrow concept of *narrative* (aka as the temporal sequentialization of two events), this is "sort of" what one can think of the *narrative dialectics of techno-scientific seeing.*

or already transmuted[8] figured world. This shifting force results in a system that does not merely repeat endlessly; it evolves, the system changing in response to minor perturbations of both negative and positive feedback. An Albanian sworn virgin who tells another man to not be so harsh to a woman, a soldier who out of necessity innovates a way of exploiting weights to scale obstacles, a paladin defending comrades who is found to be controlled by a female—these cyborgs stimulate a social precession both through their awareness and their activities. Their mere presence demonstrates that accepted answers to important questions are insufficient, and their actions show that the questions themselves may be lacking. This could be seen as a representation of the more dangerous adversary described by Shapin and Schaffer—the one "who disputed the [experimental] game" and not just the moves within it, thus putting the entire experimental community's existence at risk (Shapin and Schaffer 1985: 336).

The ideal of an objective "modest witness" can be confounded by something as innocuous as a glass that is half-full/half-empty: by forcing previously disinterested parties to squabble amongst themselves as they try to describe the same truth before them, the limited objectivity of semiotics and observations is highlighted unforgivingly. But that is not the point; it is not enough to simply critique and confuddle.[9] Without a new candidate to stand for us in our laboratories and our public interactions, we would experience a *persona vacuum*. There must be an advance in agency that better meets both pre-existing duties and the ever-changing social landscape. The modest cyborg is not the only answer, and this entity cannot fulfill all possible needs, but this is a start.

Conclusion: Cui bono?

To invert Susan Leigh Star's inversion, that it "is both more analytically interesting and politically just to begin with the question *cui bono*, than to begin with a celebration of the fact of human/non-human mingling," (Haraway 1997:38) we ask that question now. Who benefits from this new entity?

Actors and stakeholders in the citizen science movement would benefit from this conceptualization. A limitation of existing frameworks for evaluating scientific activity is that they rely on boundaries of expertise and exclusion to delineate science, making identity and not activity the distinguishing factor. It is not that citizen scientists fail to meet the standards of the modest witness, but that the modest witness fails to witness modestly yet is kept because there is

---

8   *Pace* the many options of established disciplinary *vernaculars*, we prefer to work with the alchemical idea of transmutation or, alternatively, the concept of "re-assembly" from *semantic agency theory* (SAT) (Stingl 2011a)

9   "Confuddle" is a neologism brazenly combining "confuse" and "befuddle"—a cyborg word for a cyborg article. See: http://nws.merriam-webster.com/opendictionary/newword_search.php

no viable alternative. This failure, however, does not have equitable impacts; expert scientists who inadequately meet the modest witness standard are simply dismissed as "bad scientists" (closing ranks to preserve the legitimacy of the identity) but that standard is used as a way to exclude non-experts from participating or from even being recognized as "real" scientists. A modest cyborg presents an alternative, potentially postcolonial and epistemically disobedient (cf. Go 2013, Harding 2008, Mignolo 2012) way of framing participants in the activity of science, recognizing that many configurations of people, cultures, and technology are both possible and valuable. The boundaries of expertise and culture become leverage points rather than polarizing filters; this diffraction splits the combined scientific activities into a spectrum of diverse perspectives.

The modest cyborg can also return to the cradle in online interactions—it is not just the static online environment that birthed this entity, but the interactions between users through the online environment that breathed life into the cyborg. Modesty is relative; only when you know you can remove a garment can virtue be attributed to keeping it on. In online interactions, however, this can be reversed, with the cyborg realizing that s/he can choose to wantonly cling to skins of gender, ethnicity, class, or other socially constructed costumes from the offline world, or that s/he can quietly set those aside as unnecessary and distracting. Society as a whole stands to benefit, not just from individual enlightenment, but also from tangible advances in a legal system that shamefully lags behind communications technology. Intellectual property, privacy rights, and user interactions all would benefit from more appropriate conceptual framings originating from online interactions rather than from offline limitations.

In fact, science itself benefits because the modest witness, because of its conception in the Western colonial matrix of power, will soon be (or has always been) unable to even superficially fulfill its role as purveyor of objectivity, whether perceived or actual. Without this modest witness singing from the omniscient and omnipresent risers, scientific inquiry could be silenced if no one steps up to carry the melody. There is also the problem of relying on a single voice and a single perspective to build Truth. Biological monocultures are doomed to failure and collapse; how could one expect a social ecology to be any different? Any kind of monocultures both results from and leads (back) to a *monoculture of the mind* (Shiva 1993), which—in a tragicomedy of the commons—will eventually always destroy itself.

In her/his act of epistemic disobedience, our multi-faceted modest cyborg not only brings a voice, but invites others to sing as well. Rather than producing a woeful monotone, this new chorus can create a chord, at once harmonious and dynamically dissonant, that celebrates diversity and transformation. This chorus is within reach, because in the end, we are all potential modest cyborgs waiting to speak.

Chapter 4

Sibling Saviors: A Hohfeldian Critique of Sibling Donor Conception

Sabrina M. Weiss

Reflection

In this chapter, we want to address the ethical consequences of applying techno-scientific practices. Many situations that are examined through lenses of medical bioethics and the techno-scientific practices in which they are embedded, problematized by, or produced from, also come tightly wound in many layers. These layers include concepts of figured worlds, tacit scientific assumptions and ethico-scientific paradigms wherein the welfare regimes and moral stakes of many distinct parties overlap and are turned into explicit claims.

Conventional bioethical thought often glosses over important considerations while leaping ahead to summary thinking about "substituted judgment" and "best interests standards," which often inform the ethico-legal dimension through healthcare administrative "best practices." Meanwhile, these discussions occur without actually digging deeper into how those standards are actually constructed, how they would have to be adapted to serve individual circumstances, or how they would account for novelty factors and inventive progress. One such instance, which challenges best ethico-legal practices, occurs when a family chooses to selectively implant an immuno-compatible embryo with the hopes of acquiring lifesaving donor material for an existing sick child. We can ask whether the interests of the potential life are outweighed by the interests of existing life and familial concerns. This is an occasion where the history of scientific progress and the history of legal shortcomings affect juvenile rights directly, and this is an issue that is presented and discussed here as befits the problems of *techno-scientific world-making*. We will show how the application of the so-called Hohfeldian analysis (developed by jurisprudential scholar Wesley Hohfeld) offers a system of practices built on the notion of "precise terms" for identifying and distinguishing different types of rights while framing them in relational pairings. This approach clarifies the debate, compensates for lacunae in legal precedent, and upholds important relational considerations favored by critical bioethics approaches, like feminist bioethics, thus giving due attention to many facets of concern in this issue.

Introduction

In the novel *My Sister's Keeper* by Jodi Picoult (2004), the desperate parents of a child sick with leukemia selectively conceived another child to use as a lifesaving blood and marrow donor. Over the course of the book, the reader becomes acquainted with the perspectives of the main and peripheral characters in the story as each takes a turn presenting a part of the story from their point of view. Besides providing a deeply touching story that is even more ethically relevant today than when it was first published, Picoult demonstrates narratively that this issue cannot be addressed lightly nor simply, and that an acceptable understanding cannot be achieved unless all parties involved are recognized as having interests and rights.

While issues such as stem cell research and abortion have been subjects of contentious debates in the United States, and of reasoned policy discussions nearly everywhere else, the issue of conceiving a child with the intent of using her/him as a donor has not been adequately addressed by medical ethicists, bioethicists, or by American or any other society as a whole. When attempts have been made, the vital nuances of the issue have been brushed aside under the rugs of "informed consent," "substituted judgment," and the "best interests standard" without differentiating the types of rights or dynamics of relationships inherent to this issue (Dwyer and Vig 1995). As with stem cell research, described by Glenn McGee as a "kitchen sink" problem, the practice of conceiving a donor sibling presents a convoluted tangle of conflicts, compounding the controversy with questions about genetic selection, the bounds of parental rights (or liberties), and even the purpose of a person's life. Philosophically, one can discern further conflicts between a creator and their creation, the use of people as means and ends, and the hazards of trying to weigh proximal and distal relationships without accounting for a human tendency to favor the apple before our eyes over a mere twinkle in them. Yet from this morass, we must establish an ethical conclusion that is internally consistent with existing standards, morally justified, and appreciative of the humanity inherent in familial relationships specifically, and in societal relationships generally. It is also worth noting that even the most different philosophical positions, can agree on this: "Conflict and/or moral insecurity result from an incomplete problemanalysis" (trans. from Kersting 2005: 177).[1]

Although a rights-based analysis is appropriate to assist us with identifying that which each party is due, the heavy relational component involved requires a conscientious approach.[2] Barbara Bennett Woodhouse, in *Hidden in Plain Sight*

---

1   This friction, attributed to "incompleteness of problem-analysis," is pervasive in scholarly and scientific debates concerning world-defining conceptual practices. A more (techno-scientific) practice oriented approach that engages with epistemic communities and their figured worlds can help reduce this friction, though it will never be eliminated completely. This concern for "better practices" is addressed in the conclusion.

2   We are aware of existing philosophical discussions on more general concepts of due, duty, and worth (see: Day 2002, Dworkin 1977, Habermas and O'Neill 2000, 2001;

(2008), illustrated the need for nontraditional conceptualizations of rights as pertaining to children because of their unique status as autonomous yet dependent beings. While different approaches like developmental perspectives, ecological perspectives, and needs-based/capacity-based rights were all explored, the underlying theme arose—children's rights cannot be evaluated in isolation and require a calculus that incorporates the entire rights landscape around the children in question. Feminist rights theory utilizes such an approach by emphasizing relationships over individual rights claims, but analysis lacks specific terminology[3] that could translate to use in the legal field as cases are tested in court. Therefore, in addition to recognizing relationships in families, an appropriate rights analysis method must also use precise language compatible with the legal arena.

For this end, we begin with a proposal that Hohfeldian rights analysis be used because it fulfills both standards: it is an inherently relational rights structure and it represents one of the first precise formulations of rights analysis for the legal field. By using this methodology to frame the conflicts within this issue, we can then export these constructs to other arenas for further analysis, such as philosophical or legal venues. Additionally, this method of analysis promotes relational rights discussion that is compatible with feminist theorists' concerns. Hohfeldian rights analysis then represents an optimally balanced tool for weighing rights concerns that arise in the issue of sibling donor conception.

We will first clarify the rights relationships inherent in this issue using jurisprudential scholar Wesley Hohfeld's standards for separating out different types of "rights" into Claims, Liberties, Powers, and Immunities (and their jural opposites). Then we will identify "rights complexes" that exist and interact within the scope of this issue. Next, we will explore legal precedents establishing a concept of "parental rights," discuss the lack of recognition of child rights, and present criticisms of these conventions as presented by James G. Dwyer (2006) to show why legal precedent is an insufficient tool for addressing this issue. We will demonstrate that the practice of conceiving donor siblings to save an existing child is unethical because it is inconsistent with established standards regarding rights obligations and entitlements and because it incorrectly weighs *potential* non-fundamental benefits over *certain* fundamental violations of rights. Lastly, we will address common arguments in favor of this practice using the tools of

---

Habermas, O'Neill and Manson 2007, Kersting 2005, Kersting, ed. 2000, Lister 2011, Miller 2001, Morrison 2005, Nozick 1974, Pettit 1989. 2007, Rawls 1976, 1980, 1985, 1987, 2001, Sandel 1982, Scanlon 1998, Walzer 1983). But by engaging with Hohfeld's jurisprudential framework, we hope to offer a fresh perspective that is at once compatible with traditional ethical discourses and open to inclusion of critical and postmodern perspectives. We also seek to reflect international perspectives (such as from Forst and Stingl) that do not segregate practical and applied ethics so distinctly as is done in American bioethics.

3   Re-affirming the "incompleteness trope."

Hohfeldian rights analysis to demonstrate why they are insufficient at protecting the fundamental rights of the hypothetical children who would be conceived.

A Primer on Hohfeldian Analysis

Wesley N. Hohfeld, in his seminal work, *Fundamental Legal Conceptions As Applied to Judicial Reasoning* (1918), redefined the vacillating language of rights used by the legal profession in the late 1800s and early 1900s. He criticized the conventional use of the term "rights" to conflate multiple distinct concepts: entitlements, liberties, powers to initiate and terminate relationships, immunities to exertions of power by others. In a profession so reliant on verbal precision and standard conceptualizations of relationships between entities, Hohfeld saw a distinct need to establish clearer terminologies. Thus he developed the fundamental concepts of *Jural Opposites* and *Jural Correlatives* to better describe the relational nature of rights. *Jural Opposites* were paired as opposing concepts, for example, a Duty is something that must be done, while a Privilege is something that may be done at the actor's discretion. Meanwhile, *Jural Correlatives* described "both sides" of a rights relationship; every time one of the four main concepts was held by a party, its correlative had to be held by the other party in the relationship. For example, if someone had a Right to something, then another entity had a Duty to fulfill that Right. Because our use of language has changed since this work was first written, some terms will be changed to reduce confusion and will be indicated in parentheses below, allowing us to reserve the term "rights" for more general usage.

Jural Opposites	Jural Correlatives
Right (Claim) vs. No-Right (Claim)	Right (Claim) vs. Duty
Duty vs. Privilege (Liberty)	Privilege (Liberty) vs. No-Right (Claim)
Power vs. Dis-Ability	Power vs. Liability
Immunity vs. Liability	Immunity vs. Dis-Ability

This framework is able to encompass discussions of "positive" and "negative" rights in the terms "Claims" and "Liberties," respectively, and has the added benefit of explicitly recognizing that there are other entities who must hold correlative "Duties" and "No-Claims." What is also useful about Hohfeldian analysis is that the first two pairings dealing with Claims and Duties address the substance of relationships while the latter two (Powers and Immunities) refer to the ways in which relationships are initiated or terminated. For example, when a customer walks into a store, s/he has the exclusive Power to decide whether to enter into a contract with the merchant. Meanwhile, the merchant, who must wait for the customer to make that decision, is at a Liability and will be expected to fulfill the terms of the transactional contract to exchange goods/services for currency. An

Immunity is simply the state of not being subject to another's whim regarding entrance into or exit from contractual relationships, and a Dis-Ability describes the other party's inability to affect the status of the relationship.

Hohfeldian analysis is an *effective* tool for analyzing this issue because it aids us in fulfilling the aforementioned needs for internal consistency, moral justification, and recognition of relationships. First, the clarity of language promoted by this method allows us to precisely apply standards of Constitutional rights and medical ethics to this topic. Second, by substantively distinguishing different types of rights, we are able to specifically identify violations of fundamental goods and entitlements to better weigh moral justification. Third, Hohfeldian analysis inherently recognizes that rights are relational, avoiding common critiques that rights theory is overly egoistical and isolated by conceptualizing rights as a vector rather than a scalar unit.[4]

Application of Hohfeldian Analysis

There are three types of entities to be related through rights—parents, children, and the State[5]—with two specific children recognized, the sick child and the potential donor child. Each relationship carries a collection of different types of rights, hereafter referenced as "rights complexes." The three "rights complexes" examined will be: right to reproduce, right to care, and right to self-determination.

Right to Reproduce

The "right to reproduce" is a conglomeration of multiple *Claims, Liberties, and Powers*, some held between a parent and the State, and some between the parent and a child; the concerns here are mostly *procedural*. First, a person in a free society has a Liberty to reproduce, at their option, and the State has a N*on-Claim* to interfere.[6] In some societies, a certain degree of a Claim to State assistance (Duty), whether financial or technological, is also provided to a prospective parent. This usually is accompanied by another Claim by the prospective parent to require of the State a Duty to prevent discrimination by medical service providers.

---

4   We can see a similar concern in health care, where illness and disease concepts can either be constructed in naturalizing an allopathic conceptual ontology and pathology in the form of concepts derived as "disease units" following from Cartesian dualisms or through pragmatic, relational, and integrative forms of health care practice.

5   It is curious that despite the vast literature on doctor-patient interactions and government/corporation-patient interaction, the literature and research on the actual interaction situation of the triad of doctors/governments, parents and underage patients is comparatively scarce both in number and substance.

6   See for an illustration of positive and negative liberties Berlin (1969) and Taylor (1979).

Next, a parent has the Power to initiate a relationship with a potential child by engaging in a procreative act; this is a unilateral situation because the potential child has no say in whether or not it will be conceived and birthed, so the child has only a Liability to be brought into a parent-child relationship. Once the child is born, the parent enjoys Immunity from the child attempting to terminate its relationship with the parent until the child reaches adulthood or an age at which her/his attempt to exercise a Power is taken seriously by the courts. Thus, a child for most of its life has a Dis-Ability with regards to the relationship. Meanwhile, from the moment of conception a parent (often exclusively the gestational mother) has the Power to terminate the impending relationship with the child through abortion and adoption. This imbalance of volition and capability in the procedural aspects of the parent-child relationship can be compared to the contractual relationship of marriage and emphasizes the centrality of mutual consent of the involved parties to the legality of a contract (Dwyer 1994).

While even the most oppressive marriages at least superficially allow for a consent mechanism for women to exercise some form of Power regarding marriage, a conceived child never has an opportunity to consent to being created. With the introduction of legal divorce as an option for women as well as men, wives were further empowered to terminate the marriage relationship if it became unsatisfactory by some standard; in Hohfeldian terms, wives and husbands enjoyed both a Power to end a relationship and a Liability to the other person ending it. This is used by Dwyer as an example of how a parent-child relationship exhibits unique characteristics anomalous to other relationships in free societies.

Right to Care

The concept of *care* is a multifaceted concept that has deep roots going back through Foucault to Kant and Nietzsche. Thus, it carries weight across the realms of philosophy, politics, and practice[7] and should be considered within the relationality of health, law, and family. The "right to care" complex is more substantive than procedural; a Child has a Claim to be cared for and to have basic needs met by the parent, who holds the corresponding Duty of care. This relationship is almost always kept at a private level, between the parent and child. However, an egregious failure by the parent to fulfill the Duty that is brought to the attention of the State will appoint the State as an intermediary to either enforce provision of care by the parent to the child or will sever the relationship between the parent and child to re-assign the Duty of care to another party. Because the child is reliant upon the parent to provide her/his basic needs, the child is unable to act on her/his own to enforce the fulfillment of the parent's Duty, and often cannot even petition the State directly for intervention. This lack of ability to enforce a Claim paired with the aforementioned child's unilateral

---

7   Chapter 5 incorporates this multifaceted discussion into a model of narrative empathy that engages with concepts of care.

Liability to the parent's Power to initiate and terminate a relationship traps a child in an unjust situation whereby s/he cannot demand what s/he is entitled to nor can s/he leave the relationship.

When considering the situations of the sick child and the potential donor child, one can see a new array of rights added to the complex. It is already established that a parent has a Duty to care for her/his child, so it is reasonable to assume that the parent will attempt to find a cure for the illness. However, Dwyer (1994) cautions that the existence of such a Duty does not automatically grant a right to the Duty-holder, even if that right would help her/him to fulfill said Duty. This means that a parent with a Duty of care cannot impose a Claim on a potential third party donor to fulfill their Duty even if a sick child could directly impose a Claim for their survival (via the State). However, siblings are only in a relationship because of the exercise of Power by their parents to put them into a relationship with each other; therefore a sibling should be no more Liable than a stranger. Additionally, Claims by a parent are insufficient to justify action to save her/his child since in legal discussions we do not link one person's rights to the interests of another.

Right to Self-Determination

Meanwhile, every citizen in the United States enjoys rights to self-determination provided by the Constitution, such as in the First, Sixth, Thirteenth, and Fourteenth Amendments concerning free exercise of religion, conducting one's own legal defense, prohibition of slavery, and Due Process and equal protection (Dwyer 1994). Therefore, every child, regardless of age or mental capacity, has a Claim against the State to these rights; the State has a corresponding Duty to prevent the infringement of these rights, whether indirectly through parents or directly through intervention. From this, it is apparent that in addition to a Claim by a child to receive care from a parent (or a State-appointed guardian), a child also has the Liberty to not be infringed on in certain ways, represented by a No-Claim by all others, including by her/his parents.

But in which ways should a child's self-determination be respected, when s/he is unable to evaluate or articulate preferred interests? R. Brian Howe describes fundamental rights as valid claims that should be prioritized by institutions, laws, and policies (2001) and compares them to Ronald Dworkin's "trumps" over other claims. These fundamental rights have a "weight of their own" (Rawls 1993: 32–3) separate from any instrumental value. Similarly, liberal philosophers like John Rawls and Joel Feinberg recognize "primary goods" as goods that are so basic and fundamental that their desirability is rarely argued. In cases where we do not know a person's preferences (based on culture or aesthetics), it is best to default to these as reasonably agreeable goods. John Locke's "theory of taste" likewise recognizes that tastes differ, and so it is better to avoid egregious harms and allow people to choose specific goods for themselves. These approaches are deeply codified in the liberal and libertarian discourse in the United States, where there is an emphasis on negative rights (being left alone) than on positive rights, which would need a culturally circumscribed idea of what is "good."

This demonstrates that the idea of fundamental rights not only can be described, but that they are vital in resolving situations where a person's desires are unclear (a young child) or not yet established (a potential child). And, as Dwyer noted above, when we know less about a person's desires, we have an obligation to preferentially weight tangible, universal needs over more nuanced needs. This mirrors Abraham Maslow's "Hierarchy of Needs" (1948) wherein basic physical needs like food and shelter are considered essential; until the basic needs are met, higher needs cannot be achieved. In this context, the health and integrity of one's body can be considered a fundamental right that outweighs other goods like esteem or pride at helping another, and it is reasonable for arbiters assisting a person with undifferentiated and uncommunicated interests to prefer these fundamental needs over other interests. This ought to apply in full even to a child not-yet-conceived because (assuming the pregnancy is successful) the child (with Liability to the parent's Power to reproduce) will be born and subjected to the human condition with the rest of people in society. Therefore, the potential sibling donor has a valid Claim to these fundamental rights.

Legal Precedent for Parental Rights Over Children

It is now clear that a parent attempting to fulfill her/his Duty to the sick child to care for her/him will encounter a conflict with the potential child's fundamental Claim to physical well-being and integrity. Usually, it would be appropriate to apply judicial precedent to determine the legality of such cases. But, as demonstrated by numerous Supreme Court cases, even fundamental rights have been acceptably compromised in the interests of fulfilling other standards (Dwyer 1994). The conundrum, as argued by Dwyer, is that cases generally addressing basic rights of adults (*Hodges vs. United States*,[8] *Cruzan vs. Director, Missouri Health Department*) protected the individual whose rights were being violated, but cases specifically dealing with children's rights overwhelmingly have favored parents' rights to override the rights of their children (ex. *Wisconsin vs. Yoder*). Where they did not, the parents' rights were found to be outweighed, not by the child's rights, but by the State's interests in the well-being of children (*Prince vs. Massachusetts, Jehovah's Witnesses vs. King County Hospital*) as bearing a "reasonable relation to a legitimate state purpose." Uniquely, *State vs. Miskimens*

8 In the following paragraphs, we are referring to the following cases: *Hodges vs. United States*, 203 U.S. 1; 1906; *Cruzan vs. Director, Missouri Health Department*, 497 U.S. 261; 1990; *Wisconsin vs. Yoder*, 406 U.S. 205; 1972; *Prince vs. Massachusetts*, 321 U.S. 158; 1944; *Jehovah's Witnesses vs. King County Hospital*, 390 U.S. 598; 1968; *State vs. Miskimens*, Ohio Misc.2d 43, 49, 22 O.B.R. 393, 400, 490 N.E.2d 931, 938; 1984; *State vs. Miskimens*, Ohio Misc.2d 43, 49, 22 O.B.R. 393, 400, 490 N.E.2d 931, 938; 1984.

took the groundbreaking (and non-perpetuated) step of recognizing that children had distinct interests on par with those of adults when the court invalidated the protection of religious exemption laws in charging parents whose child died from an infection that was left untreated for religious reasons. The justification was that such exemption laws violated the equal protection rights of children of religious parents under neglect laws. The court stated:

> This special protection should be guaranteed to all such children until they have *their own opportunity to make life's important decisions* for themselves upon attainment of the age of reason. After all, given the opportunity when grown up, a child may someday choose to reject the most sincerely held of his parents' religious beliefs, just as the parents on trial here have apparently grown to reject some beliefs of their parents. (*State vs. Miskimens*)

By recognizing the potential for a child to make future Free Exercise decisions, and by "[drawing] a line between exercising one's religious rights by refusing medical care for oneself, and preventing another person from receiving medical care" the *Miskimens* court ascribed both a Claim to lifesaving medical care and a Liberty to free exercise of religion to children (because this was a lower court decision, it has not been applied to federal or even state law in Ohio). However, in non-lethal cases like *In re Green*,[9] where a mother withdrew prior consent for her son to receive an operation to correct paralytic scoliosis that kept him bedridden because it would have involved a blood transfusion, such non-lethal illness was insufficient to justify State intervention. Thus, the Claim to medical care of the child in *Green* was not upheld nor recognized by the court. Even when a court decision approved State intervention, such as in *Muhlenberg Hospital vs. Patterson*,[10] it was only to prevent "grievous bodily injury," and otherwise would not violate a parent's "right … to cause their children to suffer avoidable illness of bodily injury, so long as that injury is not 'grievous'"(Dwyer 1994). When compared to laws affecting adults, there is no comparable allowance for injury that is not for self-defense reasons, yet an exemption has consistently been made for parents taking actions that inflict harm on their children.

It is apparent that turning to legal precedent regarding parental rights over children is problematic. There is a distinct lack of consistency among rulings and there is an overall lack of recognition for children's rights (Claims, Liberties, Powers, Immunities) distinct from those of their parents or from the interests of the State. Because of these two weaknesses, legal precedent alone is insufficient for justifying the violation of a child's rights to fulfill another's interests. However, in both of these areas, Hohfeldian analysis demonstrates promising advantages over existing legal precedent.

---

9   *In re Green*, 369 U.S. 689; 1962

10   *Muhlenberg Hospital vs. Patterson*, 320 A.2d 518, NJ Sup. C., Law D, Atl Report 518–21; 1974.

Addressing Arguments for Donor Sibling Conception

Three common arguments support donor sibling conception, based roughly on: necessity, negligibility of harms, and benefit for the child. These will each be addressed using the framework of Hohfeldian analysis to bolster rights and relationship analysis.

1) Obtaining a Donor Match is the Only/Most Effective Way to Save the Sick Child's Life.

Although many parents would first offer themselves as donors to save their child (which would fulfill their Duty of Care to their child), it is often the case that they are not good donor matches. Because closer matches have a higher success rate and a lower rejection rate, the odds of finding a stronger donor match must be weighed against the odds of rejection from a weaker match. However, urgent need is insufficient to negate a Claim to a fundamental interest, and as explained earlier, an onerous Duty does not automatically create a Claim by the Duty-holder upon another to enable her/him to fulfill the Duty (Dwyer 1994). For moral consistency, it would have to be acceptable to invoke such a Duty in people other than one's child. Consider a random adult who happens to be a perfect donor match to a sick child. Would it be acceptable for the parents of the sick child to demand that the stranger donate tissue or other bodily products to save the child's life, thus imposing a Duty onto her/him to enable the parents to fulfill their own Duty of care? Most would agree that this would be unacceptable because the medical community respects bodily integrity and federal law outright bans human trafficking, no matter the need; historical violations of ethics like Nuremberg were considered unacceptable for this very reason. The only difference between one's child and a complete stranger is the nature of the relationship between parent and child; this will be addressed further in Argument 3.

2) Harvesting Replenishable Bodily Materials Presents Negligible Harm and Violation of Bodily Integrity Compared to Harvesting of Organs and Therefore Does Not Harm the Donor Sibling Significantly.

While medical ethicists often look to a standard of harms to determine acceptability of an action, this issue is not fundamentally a medical issue but a legal and moral issue because it concerns not the treatment itself, but the method in which it is obtained. Therefore, a rights standard is more appropriate than a harms standard, in which case, each child individually has a vital Claim to care and health, as provided by the parent; this Claim is upheld independently of other parallel Duties a parent may have to other children. As already established, neither child has an enforceable Claim upon the other child unless they have entered into a relationship by their consent.

Even if medical ethics were used, the donor sibling would be unable to consent to the procedures involved in donation; although they are relatively low risk, there are still potential health dangers, both short and long term. Although "substituted judgment" and "best interests standards" have been applied in cases where physicians needed to decide whether to treat a child who could not consent, these were always for the child-patient who would potentially suffer greater harms through non-intervention than they would through intervention. This standard would not be used by physicians for a third party who was not the patient in question because it would violate the second child's Claim to have a physician look out for her/his medical interests exclusively, a Claim to which every patient is entitled. Dwyer also notes that such ethical standards have been used as rationalizations after the fact, rather than as part of the process determining the best course of action (1995).

Third, we can look to the general societal standards for assigning Duties of care or assistance to unassociated people. Generally, while "Good Samaritan" acts are applauded, they are seldom required, even when involving replenishable commodities like blood or bone marrow; therefore people have a Liberty to help another if they so chose, but never a Duty (barring a Duty imposed by a relationship they have entered into). Once again, we come to the question of the parent-child or sibling relationship as the determining factor for supporters.

3) The Child Will Experience a Benefit at Having Saved a Sibling

Although an already existing child could derive an emotional benefit from saving a sibling with whom s/he had a relationship, this is not so simple for a donor sibling who does not yet exist. The sticky issue of competing claims must be addressed, but it is complicated because of the temporal aspect. The problem with conceiving a child with the intent of using her/him as a donor is that the relationship does not yet exist, and any Claims for that child are "merely" potential or impending. One tendency of human nature is to more heavily weight immediate or existing concerns preferentially over potential or future concerns; we see such a bias in everything from global warming controversies to an inability of many to "save money for a rainy day." With this in mind, is it reasonable to expect a parent, or any person, to consistently give appropriate weight to a future Claim before it becomes reality? This question must be answered because any benefit a child could enjoy is completely contingent upon her/his being brought fully into existence, at which point all Claims appropriate to people would apply.

Once the child was born, we would be forced to compare possible emotional/ esteem benefits to guaranteed violations of bodily integrity, established earlier as a Claim to a "fundamental right" that must serve as a default when expressed interests or preferences are unavailable. Note that this potential emotional benefit is only possible if the child actually had the "appropriate" serotype to be useful; the implied "admission fee" that must be "paid" to a third party is problematic in both a Hohfeldian analysis and in a society that believes in "equal protections." In the event that the child was not a donor match or for some other reason (ex. hemophilia) could not donate, the impacts on her/his relationships would be damaging; it would be even more tragic should the child to learn of her/his "failure" to save the sibling. Additionally, it would be presumptuous to predict the dynamics of any familial relationship before a child was even conceived; how many conflicts are there normally within families between siblings, in-laws, and others? Recall that the child has no Power to initiate or terminate a relationship with the family once it is born, only Liability to the parents' Power. Therefore, it is presumptuous to applaud any closeness that could be gained from a "savior sibling's" sacrifice since there was no other choice in the matter on the donor's part.

Conclusion

It is a tragic circumstance when a loved one, especially a child to which one owes a Duty of care, is dying and a cure does not yet exist or is not available. And it would likely be forgivable for parents in such a situation to choose the path of conceiving a donor sibling in the desperate hope of saving their existing child; in a majority of cases it would be safe to rely on the parents' love to fulfill their Duties to both children. But it is not the place for physicians to forget their Duty as an advocate of health for every patient they treat, whether donor or recipient. Legal jurists too must reconsider the methods used to weigh difficult cases like these to ensure that no one's rightful Claims are ignored and that all people, including potential people and children, are not trapped in unjust situations where they have neither Claims respected nor Power granted to leave unfair arrangements. For a Duty to predicate all Claims and even the existence of a person is to risk allowing people to become mere instruments. The only way to safeguard against such a violation is to utilize precise and consistent analytical methods from the start to provide common, stable ground from which to derive ethical solutions to the problems that plague us all.

This chapter builds on our critique of the idea of "pure" knowledges—both scientific and philosophical—and highlights difficulties that can arise from oversimplifying the discussion. As with the sociology of mathematics, this engagement with an issue that is not given due consideration within a complex ethico-legal ecology illustrates how this knowledge-practice is embedded in techno-scientific figured worlds spanning history, relationships, and legal institutions.

Chapter 5

Local Biopolities of Solidarity, Epistemic Cultures of Empathy, and the Civil Sphere

Alexander I. Stingl

Therapies for Being "Good at Heart": More than a Socio-Medical Problem.

In this chapter we argue that cultures of techno-science can exist in democratic societies but require a realistic, functional, social, and moderate semantic holism. In a Durkheimian account of Jeffrey Alexander's *The Civil Sphere,* we show that techno-science, scholarship and their publics can achieve democracy as a form of "epistemic multiculturalism" or integration that comprehends inclusion with the necessity of exclusion. By re-conceptualizing agency in this framework, we can resolve three basic problems concerning the integration of epistemic cultures: intersubjectivity of the Other, achievement of solidarity, and collective versus individual agency in networks. We offer narrative empathy as an effective way to theorize and account for these problems, and to distinguish it from story or fiction; we define "narrative" in a narrow fashion that conceptualizes agency as a trinary form of decision-making. Instead of accepting fragmentation through impermeable boundaries between epistemic communities, we celebrate a democracy of heterogeneous communities of techno-scientific practitioners and their publics (stock- and stake-holders) wherein integration is achieved through a civics-based form of solidarity. *ScienceCraft* and *ScienceCivics* are two sides of the same, narrative coin, wherein the decision-making practitioner emerges as the third or the parasite that integrates.

Democratic societies are not designed to have a "good heart." By this, we mean to say that citizens of a democratic society are not naturally "good at heart." But then, they are not intrinsically "bad-hearted" and democratic societies are not inherently evil either. Interestingly, what is true about the moral heart is also very often true about the heart as an organ, and it is over a life-course that hearts "become." Democratic nations, like any other society, are formed around a process of becoming, also known as boundary-formation: this is built upon two complementary sub-processes, inclusion and exclusion. These two sub-processes are value-laden tendencies that manifest through evaluative deliberations (Taylor 1999), but it is important to recognize that they are distinct and complementary, not only antagonistic.

How we deal with both as forms of *ScienceCraft* is what makes these processes interesting. Because techno-scientific practices and the political imagination are deeply intertwined and interdependent, political action is necessary and relevant: We can have neither a democratic society nor any other epistemic community without the idea that some people belong to it and others don't. How we deal and interact with those that don't, what we do in order to try and include them into our deliberations and (collective) actions, says a lot about us, and that is what in part defines "civics." Civics has recently seen an overdue revival; on the one hand, with regard to the question of how democratic life can help negotiate global–local differences (Alexander 2006) and, on the other, with the regard to the potentials for and consequences of the rise of biopower (Rose 2006). This double revival rests on the fundamental claim that despite the so-called "victory" of "Capitalism" over "Communism" and the cooptation of science by management paradigms, inequalities and social injustices have not miraculously disappeared (Brodie 2007). Science has not *managed* to solve our most pressing problems; in general, the gaps in wealth, knowledge, and power within "developed" countries and between "developed" and "un/under-developed" countries have been widening. The reasons are historic-cultural (e.g., class divisions) and contemporary (e.g., due to the effects of the *digital divide*), as expressed by Slavoj Žižek:

> One can sincerely fight to preserve the environment, defend a broader notion of intellectual property, or oppose the copyrighting of genes, without ever confronting the antagonism between the Included and the Excluded. Furthermore, one can even formulate certain aspects of these struggles in the terms of the Included being threatened by the polluting Excluded. (Žižek 2009: 98)

The need and the possibility to *empower* us as biological citizens who are constantly negotiating with various experts and specialists is the other important dimension and, as a discourse, has been underwritten by Nikolas Rose (2006), Aihwa Ong (2006), Adriana Petryana (2002), among others. Rose in particular has tried to help us navigate the unruly waters of inequalities stirred by the biomedical sciences, whereas the more general problem of knowledge economies in modern societies has been a feature of the intellectual landscape since the early 1960s. That landscape is marked by biomedical dimensions, the morals of markets, and the impacts of the digital divide.

And yet, these structural assertions cry for agency. Fritz Breithaupt's Kulturen der Empathie (2009) has created a saturated account for this need to reinvigorate the civics discourse in a lasting way. A new type of social theory seems not just called for but also possible.

Empathy

This discourse was initiated by two items that may at first seem unrelated: A recent report tells of a study offering clues that among student populations in the past three decades the ability for empathy has declined dramatically. Even if the concept of empathy used in those studies may be under-theorized and insufficient, these studies do offer ample warning of a danger afoot for civil society and beyond. This second aspect reminds us that Durkheim's positive idea of a civil religion was centered around a concept that points to medicine. Given that we live in an age of biological citizenship, it follows that if biopolitics is to be part of any contemporary democratic society it must be saturated with a concept of civics; hence, *biocivics*. At the heart of this bio-polity is, of course, the concept of care and the interaction of (*expert*) care-givers and (*lay*) care-receivers, modeled on the ideal-type of the *doctor-patient* interaction. Recently, the old conflict between phenomenological and operational diagnostics in "health and care" settings has flared up again, eliciting the response that, unfortunately, phenomenology is as good dead with the rise of Neo-Kraepelianism. The switch from educating students—in the health and mental health area in particular—to *merely* training them accounts for this change. At the heart of Jaspers's integrative philosophy of medicine (and similar models, including complementary medicine [CAM]), lie both empathy and a narrative dimension. Therefore, any form of biocivics that can be taken seriously and accounts for the issues raised by Alexander and Breithaupt will have to understand that medicine has a crucial function and potential as a "form of civics." The (narrative) empathic skills of students and practitioners in care-giving, as well as of intermediary agents in relevant lay-expert interactions, must be nourished. Doctors in particular must be better embedded in communities to help form effective civic networks that strengthen democratic values, promote social justice, and mitigate the harms of the digital divide in the age of biological citizenship. Civil society as a project (following Alexander 2006) begins with biocivics and medicine "at its heart"—a cardio-practice, so to speak—and with the practice of *narrative* empathy.

On the Premises of Social and Semantic Holism

Our discourse here mostly follows the premises of a semantic holism that is qualified as moderate (Seel 2002) and social (Pettit 1993), and which has been delineated through a set of modifications by Michael Esfeld (2001, 2002, 2008, 2012, with Lam 2008a, 2010). This forms an interesting cornerstone for a theory of knowledge that accounts for semantic agency in meaningful practices. Abstracting from this, the consequences for epistemic communities of physicists and related natural sciences can be generalized for epistemic communities of scholars, scientists, and their respective publics, as illustrated through *local biopolities* of biomedical and care communities. The concept of *polity*, just like the concept of *community*, is not sufficiently reflected in contemporary discourse despite great potential to add interactional and relationist layers to the dialogical models of Ego/Alter, Ego/"Generalized Other" or Individual/Society. To enable integrative techno-scientific practices and overcome the limitations of First/Second and First/Third binaries, we must accept that we need a different order of model that introduces *thirdness* (Breithaupt 2012a, Esposito 2012, Lindemann 2012, Mignolo 2012), embraces the *parasite* (Serres 2007), and complements—rather than replaces—it with a "holism from below" (Greshoff 2008, Lindemann 2005, Schützeichel 2009).

Social holism, in general, is based on the conceptual reconstruction of the practice of "following rules": Rule following is constituted as a social practice, and it is claimed that, in turn, social practices constitute *meaning* (Esfeld 2002: 108). Therefore, outside of practice, no action is meaningful. Subsequently, any form of agency has to be considered relational and, accordingly, semantic agency.

The personal faculty that facilitates this aspect of agency is judgment, realized in the Kantian sense as predicating the rule in practice, and as a power that must be continually exercised, practiced and trained, often by "engaging in exemplary action." As an ability, practice (of this kind) is defined as the "knowledge *how*." Persons, if they are to have beliefs of any kind, are ontologically dependent on other persons because meaning cannot be constituted mentally, but only in practice. This dependency leads to the conviction that two such persons form a social and, probably, linguistic community.[1] At the same time, this "analytical" discourse is also just a corollary of Kantian philosophy, after a fashion (Hanna 2001).

Social Holism now reconstructs how social practices provide the regulation of meaning (as comes into play in our beliefs) according to a set of steps that leave certain areas problematic once exposed to the consequences of knowledge economies in the age of biological citizenship, of narrative empathy, and of the civil sphere. These steps are as follow (as explicated originally in: Esfeld 2002: 109–14; 127–32):

1. Although every definite set of examples can be expanded and continued in an infinite number of ways, every finite being is generally dispositioned to conclude such a set in a specific way.
2. Persons sharing a common biological equipment and a physical environment will not have "bizarrely" different dispositions.
3. Dispositions of people with a common biological equipment and shared physical environment include a disposition to coordinate one's own behavior, at least partially, with that of others (Second Order Disposition).
4. The disposition of minimally partial coordination leads to persons reacting to their actions by mutually sanctioning them through affirmations or restraint.

---

1 "Semantic" here refers to meaning, not just as linguistic/lexic meaning, but as any kind of meaningful act that can also be deictic, physical, etc. (cf: Shalin 2007, Stingl 2011a,b). We note that in collaboration with the late John Schumacher, Restivo has generalized the Einstein clock synchronization thought experiment in a direction that makes the triad the fundamental unit of society rather than the dyad. Their work exists in an unpublished draft ms. See the discussion on Breithaupt and triads that in the next section.

5. Sanctions are a means to generate the conditions that allow persons to achieve a consensus in the way that a given set of examples is to be concluded.
6. As a consequence of the process that generates the conditions for such a consensus, a rule can be considered the element that constitutes the convergence of persons to conclude a given set of examples.

As a necessary prerequisite for this process, Esfeld introduces as a (bizarrely Habermasian-like) criterion of transparency, the following demand: The process must be transparent for all the participants whether or not their reactions toward the environment, at any particular time, converge or not.

> 7a. Since pure social holism would be inherently normative and deterministic—respectively, persons would be 'judgmental dopes'—a lesson from inferential or belief holism must be introduced as a prerequisite so that people *can* follow different and differing rules:[2] Just knowing how to follow a rule does not tell us anything about the content of the rule. Therefore we need to accept that ...
>
> 7b. ... mutual sanctioning as a form of judging one another's actions can determine the conceptual content of a rule *if* the content is determined for an open number of further rules; further rules comprise the *inferential context* for the beliefs that are constituted by following the rule currently in question.

Respectively, conceptual belief is based in social interaction, which would relieve us from "translating" another person's beliefs, for these would be determined and individuated by "the same external factor." In the most radical suggestion, translation of any kind of conceptual content is defined as a *melting* of the communal contingencies (of two or more communities) into one community practice that can be successful or not (Esfeld 2002: 131; Lance/O'Leary-Hawthorne 1997). In reality, *melting* can take many forms in various contexts that altogether shape public discourse. Esfeld gives us a mere description of "knowing how" without an actual account of agency. This is the very problem that Breithaupt's empathy can help us resolve.

As for the question of the person-environment relation and the idea of "common biological equipment," in the age in which we understand ourselves and are treated by the "governing and disciplinary institutions" as neurochemical selves, the fact remains that the idea of the "same" biological equipment has become a) questionable and b) subject of construction through social interactions

---

2   The good old question (see also: Reich 2010) whether Parsons' social theory leads to constructing persons as "judgmental dopes" hinges on what type of holism Parsons would have to followed based on his construction of agency. Parsons' persons are anything but judgmental dopes. But then again, those critics who stick to the dopey-account would probably like to declare all "Kantian persons" judgmental dopes, too.

itself.[3] Hence, the inclusion of this question shows how dependent any holistic account is on satisfying the questions raised by biology and, more importantly, biocivics and biopolitics.

The dimension of "the body and its expressions" is under-theorized in any merely dia-logical account. But the relationship between one person and another includes the body and its expressions with necessity (Shusterman 1997, 2008). Therefore, any interactional situation is also dia-*physical* and trinary, even when the body is not physically present, such as in internet communication (whether the criterion of reacting to the environment "at the same time" is satisfied by written correspondence is not even accounted for in this list of issues). The non-presence and presence of the interacting bodies changes the situation and environment significantly.

Fritz Breithaupt makes a good case that interactions, involving the kind of understanding of another person in his/her following or not following a rule, are triadic and, based on a person making a judgment on two persons' opposing positions, embedded in narratives (see below). We would like to assume, respectively, that inferential contexts and narratives are cognitively and structurally of the same design. While we do not yet fully know what an unfolded triadic social-moderate holism would look like, Pettit's and Esfeld's model can easily serve as a basis for such a theory, which would prove to be compatible with a new attitude in the philosophy of science studies, one we seek to capture in the concept of *ScienceCraft*.

Civil Sphere and the Need for Agency

Modes of Incorporation

The *civil sphere* is "a new 'social fact'" (Alexander 2007: 10) that represents a new concept of what civil society is—a project.[4] This project, to paraphrase Kant, is underscored by the insight that we do not live in *civil* times but in *civilizing* times; that this kind of civilizing can go either way emphasizes both the danger and the ambiguity we face. The civil sphere and its civil society are no guarantee of a better state, a better integration of out-groups, or the inclusion of and justice for *all*. Instead, the civil sphere could also constitute a more narcissistic state, a shedding of out-groups through exclusion, and greater injustice towards these out-groups. This is not a moral judgment, for the civil sphere is, in a certain sense, beyond (or before) good and evil even though it tries to do its best; the civil sphere is a *capacitator* rather than an *incapacitator*.

---

3   Chapters 3 and 4 have engaged with this concern by discussing parallels between social and material technologies in the context of cyborg theory.

4   Critical appraisals by Hall, Kivisto, and Griffin in the same journal issue respond to Alexander.

Integration requires both inclusion and exclusion. To integrate or become integrated means to go through mechanisms of inclusion and exclusion in parallel—mechanisms that call on *solidarity* and that filter difference adequately. Alexander (2001, 2006) does not theorize solidarity on the micro-level of social agency so we must turn to Breithaupt's narrative empathy.

In the current discursive climate, Alexander asserts that political conservatives and radical multiculturalists each offer partially correct judgments. The conservative critics are right that there is already a civil sphere in the democratic and democratizing nations. The radicals are right that contemporary civil societies are fragmented and fractured (Alexander 2006: 402). The liberating potentials of contemporary life are the consequence of the tension between the realities and ideals of the civil society. Overcoming this tension on the group level has two aspects: one on the level of agency and one on the level of structure. In ideal typical terms, following Alexander (2001a), civil societies have three options or *modes of incorporation* to resolve the tension through a parallel running inclusion-exclusion process: (a) assimilation, (b) hyphenation, and (c) multiculturalism

Assimilation—For out-groups or out-group members to enter civil life, assimilation requires them to let go of their differing primordial identities. This is, according to Alexander, a purifying process that has little to do with the qualities that constituted these foreign identities, for these are in themselves left untouched. The persons that carry them must let go, shedding them entirely in public life. To enter the public arena of group A, a member of group B must let go of all the qualities b_1 to b_n, that constitute membership in group B and adopt (possibly all) the qualities a_1 to a_n, that constitute membership in group A. Due to the public nature of assimilation, the assimilated retain instabilities, even if reified.

Hyphenation, which historically led to the emergence of ethnicity as an idea in itself, is a more creative process. Helping to "tolerate" foreign identities and allow for fluidity or interchange of "primordial qualities" (inclusive of bridges between the private and the public sphere), leading to *hybrid discourses*. However, and this is important to stress, Alexander asserts that this does not render outsider and core primordial identities equal. The core identity remains supreme and retains levels of stigmatization of outsiders. This, too, results in instabilities.

Multiculturalism, in Alexander's perspective, is not the all-inclusive type of category that some of the writers he calls "radical multiculturalists" consider or, rather, want it to be. This is not a happy-go-lightly and anything-goes category either, for it still is a *purifying* process of some kind. However, it is not individual people that are purified, it is the primordial categories that are being purified. In the process, differences are turned into experiences that can be common, thereby changing the direction of particularity and universality. Whereas assimilation and hyphenation indicated a universalization of the particular, in the multicultural mode the universal is particularized and diversity is achieved through performance. This suggests, however, that the purification of the out-group qualities means that they must be opened and re-created (hence, purified) in an accessible fashion for the core-group to be able to participate. In other words, the primordial qualities of the out-group must be made intelligible for the core-group while remaining different from core-group qualities. This is what Alexander seems to mean by "achievement."

Here we must ask: Who achieves and performs? Who or what is incorporated? Who is *solidary* and thus executes *solidary* action? Who incorporates? Do we deal with actors and/or agents on the level of the civil sphere who are acting in these modes of incorporation? Our goal is not simply to understand Alexander's ideas but to apply them pragmatically both in political action and in the creation of a better social theory. To accomplish this we need some conceptual precision.

Generally, in social theories we usually deal with one of two types of acting entities to begin with: individuals and collectivities. There is still some question about whether collectivities are merely aggregates of individuals or collective actors. Talcott Parsons, although he did not discuss the matter explicitly very often, understood collectivities to be comprised of and dependent on individuals. For Parsons, only individuals could be agentic, meaning that only individuals could act on and express intentionality. While there could be collective intentionality—in the way Searle would speak about it today—even in Parsonian theorizing, this is derived from Durkheim and—*minus the evolutionary perspective* — Alexander is to some extent also indebted to Durkheim.

Therefore, we might be tempted to ask: Are Alexander's groups to be considered aggregates of actors or truly *collective* agents? However, this is not necessarily the question we seek to answer since we are not averse to discussions of *third*, relationist options. We are specifically open to such options because we seek to avoid the very fallacies we outline at the end of Chapter 2.[5] The question that makes our little chapter tick, then, is: What level of agency makes an expression of a symbolic representation *effective in practice*? Reformulated with special regard to civil society and the civil sphere, the question becomes: Does this type of agency *in practices* lie with individuals, with groups, or with both? And for civil society, what about institutionalized organizations of democratic society?

While Alexander blurs those lines a bit, we will have to resolve the issue of how individuals and groups interact within an environment of organizations (see Stingl 2011). The solution we suggest is based on the notion of a "meso-level" or intermediate type of entity: civic networks account for the plurality of possible roles and memberships of an individual while transgressing the privacy/public distinction (to some degree). It is this partial transgression that enables an individual to function in those capacities necessary to express collective intentionalities effectively. Enabling agency in civil societies means that solidarity, in this case,

---

5　The fallacies were named and defined as the transcendental, subscendental, private worlds, internal life, psychologistic, and eternal worlds fallacies. Each could also be recast in the terms of individualism, atomism, strict holism, and so on. Instead, we counter-propose a pragmatism that is forged from a relationist, structure-proceduralist realism that construes of (techno-scientific) practices in a moderate holism.

is really a form of empathy that provides a solution for "three problems" in social theory: integration, intersubjectivity, and group agency.

Problem 1: Durkheim

Durkheim is credited with developing a proto-theory concerning the integration of out-groups (Kenny 2010) and for introducing a "cult of humanity" as a regulatory alternative to constitutive solidarity (Dew 2007), which can be understood as a new civic religion focused on medicine and public health. Although Durkheim's moral theory, dubbed "belonging is believing," was subjected to criticism by some commentators, these readers failed to appreciate the situatedness of these ideas as contemporary to nineteenth-century science.

Agents as inhabitants of complex social systems are "exposed to alternative moral lifestyles," which cause them to experience dissonance at the emotive, cognitive, and performative levels (Kenny 2010: 217). Coping competencies come in two forms: sophisticated and unsophisticated reflexivity. The agents derive "social coping competencies" to deal with these dissonances, and "alternative moral lifestyles" can be translated into *out-groups*, since "where a group is formed, a moral discipline is formed, too" (Kenny 2010: 218, quoting Durkheim).

This leads Kenny to derive the following corollaries for what, with Durkheim, he understands as *moral spheres*: moral orientations are the context for ritual solidarity; solidarities are plastic and spontaneous; plural moral affiliations define the healthy state; and each solidarity derives its morality from meaningful symbolic action. As a key component of bridging moral spheres, Kenny (2010: 234) claims that reflexivity is reasoned engagement, or in a Kantian context, the faculty of reason. That said, Kenny adopts Alexander's modes of incorporation but without recognizing that there is an inherent need for exclusive forces. We find here a correlation between Kenny's "islands of moral immunity, Alexander's "common humanity worthy of civil respect," and Durkheim's "cult of humanity."

This lack of realism for the requirement of exclusion and lack of differentiation between inclusion and integration that we sense in Kenny's reading of Durkheim is the reason why we feel that his discussion, while impressive, does not "overtake" Alexander's account and instead remains behind. But Kenny nevertheless points to the lack of saturation in Alexander's recent opus magnum on the micro or agentic front: The agent in the civil sphere is moving between moral spheres, that is, between *solidarities*, and this dynamic is inspired by "motivations." Practices constitute the civil sphere, but what constitutes practices? Practices—even though they must be reconcilable or abandonable, at least symbolically—must be the *givens* in and for a civil society, its civil sphere, and its theory. At the end of the day, every agent and every actor is entangled in a process of decision-making when "deciding for a practice." And while symbolically represented, embedded in culture, and wrapped into internal environments, there is a form of mechanism for this decision-making, a form that has two components: a reflective or enabling (here: cognitive) one, and a constitutive or constraining (here: institutional) one: The constitutive mechanism that regulates this process is all about the *governmentality* of knowledge and truth itself; it is also the one with which we shall not busy ourselves here. The reflective and cognitive aspect is what we will deal with and we identify this mechanism (of decision-making) as narrative empathy.

Problem 2: Intersubjectivity

The problem of intersubjectivity, in one breath, is the problem of how one person can understand another person, even though they cannot "look" into one another's minds. And it is this problem that leads us to seek a resolution through concept of empathy distinct from that of the phenomenological tradition (Zahavi 2010a).

Historically, the concept of *intersubjectivity* is a name for a *residual category* that popped up in Early Pragmatism and which later generations have come to *reify* and treat as a natural kind or reject altogether. In *Civil Sphere*, Alexander avoids the problem in two ways, (a) explicitly in his differentiation between public and private, and (b) implicitly, a priori, and, more profoundly, in the undertheorization of agency. Rather than attempt to amend or criticize, we seek to show that the *Civil Sphere* holds the promise of a new pragmatic social theory when it is recast along with discussions of agency and knowledge

Wendelin Reich identifies "three problems" with intersubjectivity (2010): 1) the "philosophy of other minds," 2) the idea of "shared understanding," and 3) "double contingency." The first problem, "philosophy of other minds," derives from Descartes' notion that other people are subjects whose inner states may or may not be different from my own *(first person)*.[6] In the history of (natural) philosophy arguably just prior to Descartes, the purpose of someone else's actions emerged as a *problem* for philosophy. Reich claims that this problem initially was brushed away by Idealist philosophers via the supposition of some commonly shared faculty or transcendental conception like Reason, per Kant, or Spirit, per Hegel. Only the Anglo-American philosophers of the later nineteenth and early twentieth century would then "make it explicit"—"it" being the other-minds-problem—and seek to find out how to justify belief in other minds (Reich 2010: 43). Accordingly, Reich summarizes that Bertrand Russell went over to another idea that was the notion of a "simulation" or "analogy of self and others," followed by Wittgenstein, Martin Hollis (1977), Michael Levin (1984), and which led up to recent cognitive science.

Husserl's work generates a phenomenological account of intersubjectivity as "shared understanding," the second problem identified by Reich. Husserlian *understanding* is grounded in *empathy (Einfühlung)*; by contrast, Breithaupt and Keen (see below) introduce a less contemplative analytics of empathy than that of the phenomenologists. Sartre's distinction between deterministic causality and indeterministic selectivity raises the issue of self-imposed contingency in the

---

6   Here and in the following paragraphs, the agent in the civil sphere is moving between moral spheres. Recall the fallacies we pointed out earlier (Chapter 2). And on the problem of intersubjectivity as a problem in the sociology of knowledge, see Restivo 1994: 181ff.

observation of others. Husserl discusses events and objects as indices for acquiring knowledge of events and of co-interactants. First- and second-order knowledge, respectively, is distributed across social encounters (Reich 2010: 50).

With Schuetz's supposition of in-order-to motives, Reich's interactants seem to take the indexical events to be acts of their co-interactants, and a co-interactant's bringing-about" of an act is, consequently, identified as an intention. This is riddled with problems. First, Reich would have been well-advised to differentiate the group from individual actors since the level of agency nominally pertains to both, as the "in-order-to motive as intention" is an agency property that is read from indexicals. Without a proper semiotics, this would mean that, at the same time, all acting entities would have to be assigned the same properties of agents; therefore, groups would need to have subjectivity.

There is also the danger of identifying motive with motivation.

Finally, Reich claims that in conflating the phenomenological and analytical problems, we would get to a "big picture of the achievement of intersubjectivity in social interaction." After that statement, a suspicion emerges: Whereas Alexander sees achievement as a conceptual prerequisite in processing or interacting in the mode of multiculturalism, Reich unequivocally identifies solidarity with intersubjectivity. This position, that sociologists ought to complacently leave the problem to evolutionary anthropologists, would be criticized as contemplativism by Parsons. To us, there is a deeper concern: Reich's position represents a dangerous road to fatalism and defeatism paved by the creation of impermeable boundaries between scholarly and scientific disciplines. With the suggestion that understanding someone's *others* was the sole basis for and sole demand of solidarity, the dimension of social conflict would either have to cease to exist or be considered inherently irrational and thus deemed unresolvable from a sociological or any inter/multi-/trans-disciplinary perspective. All that sociologists would then be able and allowed to do would be to take notice of any potentially conflictual development ... and then shut up about it—in the Wittgesteinian motto, that what you cannot speak about, you have to be silent about. But this is a kind of "silent *othering*": Disciplines are silently turned into *others*, and then in their *otherness* they are silenced.

Because they have very much accepted the fiction that academia in general and the so-called "soft sciences"—the humanities and the (all too) social sciences—have lost, they have created a fetish to bemoan. Žižek commented on this pessimistic defeatism thusly: "The Leftist nostalgic's attitude is: 'You see, our side is losing!' 'Which side?' 'We'll know that tomorrow, when we find out who lost!'" (2002: 53, fn.24). They fulfill this by constantly advising against careers in their field and emphasizing the hardships of scholarly life; at the same time they adopt a singular best practice that promotes the growing business of academia and keep jumping through their own hoops to progress. Such a conclusion would certainly be as unacceptable for Jeffrey Alexander as it is for us, which is why, with the inclusion of Breithaupt, we are enabled begin to create a new social theory that is decisive and interventionist on a pragmatic footing.

Reich's third problem of intersubjectivity, double contingency, focuses on Talcott Parsons and Niklas Luhmann. Parsons first formulated the problem of double contingency: It is a fact that ego and alter ego are mutually nontransparent, mutually interdependent, and thus, social systems emerge that are laid down in expectant structures that provide motivations and goals, which can either be satisfied or disappointed (1951).

Reich rebuffs the idea of a Parsonian solution by reminding readers of the criticism Parsons's conception received after its initial decades of success, criticisms such as Wrong's "oversocialization charge" (Wrong 1961) or Garfinkel stating that Parsons' actors were "judgmental dopes" (1967). Reich then argues that Parsons' actors do not attain "the interactional status of the subject" (Reich 2010: 52), invoking Sartre for comparison. Luhmann used Parsons's double contingency model but rejected Parsons's theoretical framework for his own version. Unfortunately, Reich's account of Luhmann seems somewhat off, even though he is right to assert that Luhmann was highly skeptical of the idea of intersubjectivity itself. But to Reich, both thinkers present a step towards the solution for his problem because Parsons identified the need to "convert the psychological problem of mental intransparency into a sociological problem of social coordination" and Luhmann "shows that any such coordination [...] has to be a process" (Reich 2010: 55). We must take issue with this conclusion, however, because that understanding is highly undertheorized; in other words, it would need to be properly defined by what counts as (working) knowledge and what counts as truth. Additionally, to bring this back to Alexander, we require some form of agency or interagency because the interactive dimension is fundamentally constitutive for his concept of solidarity. The "conversation" between the concepts of *intersubjectivity* and *understanding* does not resolve Reich's problem, however, because the problem of how two persons can understand one another at all was his initial question, which he tried to solve by showing how this is answered by theories that explain interaction through coordination.

Yet, even Parsons' theory is a more adequate theory of coordination and cooperation. The fact that ego cannot know for certain what alter ego will do can be explained from a Kantian premise or from the sort of bio-organic theorizing one finds for example in Barnard (1968). This was the gist of the Human Relations Movement by L.J. Henderson and Chester Barnard, and its successor, Parsons' Social Relations Movement. The act of "deciding" and the ability and willingness to cooperate is manifest in overcoming biological limitations. In epistemological terms, between tautology and regression, random abortion and stepping over the boundary is the act of achieving transcendence of that which divides us. Successful communication is not "understanding" but achieving a useful result through coordination.

Reich does not arrive at this Parsonian solution. In the end he fails to recognize the interdisciplinary ways in which the three discourses he identifies as distinct are in fact deeply intertwined and reference a shared problem. In trying to keep them separate at the start, Reich only imputes and ontologizes disciplinary boundaries.

His solution finally takes the form of distinguishing *observational understanding* from *communicative understanding*.

Observational understanding is "simply" a form of behavioral understanding achieved through the computation of the physical signals (body movements) sent by someone else; these signals are considered to be unintentional.[7] Communicative understanding relies on the participation of both an "understander" and an "understandee": the former reconstructs practical motives while the latter prevents irrelevant unseen motives from being assigned. Additionally, the understander evaluates the understandee's behavior using "socially standardized rules"; the standardization allows the understandee to anticipate their use, but it is important to note that these rules vary across cultures. This is a "socially viable strategy" because two "coordinative interactional mechanisms" that are constitutive of any interactive situation interlock with each other: "pragmatic accountability" and "sequential adjustability" (Reich 2010: 58). These can be sensibly explained as, "people who interact have memory," and "people take turns in interactions when they communicate through language," suggesting Reich is simply packing common sense with unnecessarily obtuse jargon.

He also does not take us beyond any of the theories he discusses nor does he provide us with a functioning synthesis. Indeed, it sounds like he is trying to reinvent the Habermasian wheel without doing the necessary work to round the edges off a square form. Subsequently, his conclusion that this conceptualization offers analytical clarity in distinguishing culturally universal coordinative mechanisms from culturally specific interpretative rules within cross-cultural modes of social understanding remains unsubstantiated. What we know, though, is that his trust in "happy intuitions about social life" and his "leaving other matters to evolutionary anthropologists" (Reich 2010: 60) is not sociologically viable.

However, his work helps illustrate that the *Civil Sphere* and its concept of solidarity, despite its inherent strength, can and should be saturated with a concept of understanding that accounts for the aspect of agency (or interagency) in the situation of understanding. Such a situation involves decision-making and the rendering of decisions into practices and is not restricted to being dyadic, as is implicitly suggested by Reich and the phenomenologists, but is at the very least *triadic*, as we will see. In looking at Reich, we make apparent the shortcomings of numerous contemporary social theories—those of German-style theorists in particular (including many followers of Schuetz, Luhmann and Habermas)—and illustrate the need for a new chapter in social theory. For example, Restivo, as early as 1974, took intersubjectivity out of the hands of the philosophers and sociologized the concept; a sociology of objectivity was the result (Restivo 1994). But here, we are more concerned with agency in a pragmatic sense and with teasing out its relevance for social practice.

---

7   Any serious theory of action or intention would have to take this apart, given the role that "unintentional" signals play in intentional actions.

Problem 3: Agency in Groups

Who or what has the capacity to act? Following this, what is agency? We often speak of groups and organizations as if they had agentic properties but offer no evidence that they do. If we were to accept that groups possess agency in a form that transcends the (aggregation of) agency of the individual members of the group, then this could also affect our overall understanding of social agency and thus affect Jeffrey Alexander's civil sphere theorem.

Philip Pettit (2009), following Dennett and Stalnaker (1984), begins his account with a discussion of "agential behavior." Different types of entities display "agential behavior," patterns that are purposive in so far as they are aimed at reliably achieving certain outcomes. The patterns comprise a system of contingent adjustments the system can make to realize these purposes, patterns understood to be representations that necessitate the situational adoption of effective means. Behavior that displays these features is *rational*.

For an entity to have agency, it must, at the very least, afford agentic behavior. Agentic behavior occurs on an *ascriptive* level, meaning that we ascribe agentic behavior to these entities by using an agentic vocabulary to implicitly ascribe intentionality and intentions. In trying to make these ascriptions explicit, we find whether or not this is a true case of agency (see also Rios comment on Pettit's discourse following Petitt 2009).

The steps we need to take are to elucidate the very purposive-representational patterns that pertain to the situation and behavioral entity, and to identify the pattern as transcontextual. We would also need to demonstrate how this behavioral frame applies to and is applied by *transparent* groups, referring to groups where all members have all the information. Finally, we would need to show that the same holds true for groups that are *intransparent*. Pettit's position seems to be that, in specific cases of decision-making at least, showing that the framework holds up for transparent groups will naturally allow us to extend it to intransparent groups. This rests on the notion that *there are* certain established mechanisms of decision-making (in Pettit's example: the *straw-vote assembly*). There is potential for a social theory to help generalize the concept of group agents or to at least make patterns and entities explicit in order to make complex theories of knowledge regimes and decision-making more accessible.

Following this, we ask: "Who are Alexander's agentic entities, and what are their behavioral agentic patterns?" To answer these questions, we must create a social theory that can encompass Alexander's discourse. Meanwhile, Alexander's *civil sphere* provides a diagnosis of established mechanisms of decision-making in the form of the *modes of incorporation* and the civil sphere itself. In complex civil societies, "group agents" are not necessarily found in organizations or institutions (social structures), but in civic networks (polities). Likewise, these civic networks represent mechanisms of decision-making. Individual agency must therefore function for individual members as representatives of these networks and of private (supposedly dyadic) interactions. We will see that the basic triadic structure of narrative empathy resolves the problem of agency on both levels and relieves Alexander's theorem of the burden that the problems of intersubjective understanding and the composition of agentic entities pose.

Accounting Reliably for the Other's Frame of Action Potentials and
Its Actualization through Decision-Making

Agentic entities are embedded in political structures and situations that enable or constrain not only their actions but also their levels of integration with one another, with the social structures, and with their environments. The civil sphere, civil society, and the cultural differences of groups represent such a case. When two groups meet in the civil sphere, they assess how far their respective frameworks, which delineate the potential expressive actions they can take, can be reconciled; this represents at its heart the discourse on justice and tolerance.

Civil society has a core group that has to deal with out-groups in just such a (simplified) situation. The act of defining a group or an actor as an out-group or member of an out-group rests on the assessment of the other's frame of action potentials and of the mechanisms of choice. An out-group mechanism of choice can restrict a wider frame of action potentials significantly and can allow for levels of integration that would not be possible based on the assessment of the frame alone.

In a simplified case, an out-group's frame, say of religious beliefs, might suggest practices of treating certain animals (e.g. butchering practices in Jewish, Hindu or Muslim communities, or using animals for laboratory experiments) in ways that do not fit in with the core-group if the practices were *expressed*. However, decision-making processes may allow for the out-group to make a choice to pursue or not pursue this practice. Therefore, mechanisms of decision-making must be paid as much attention as the frames for action potentials. Alexander's *Civil Sphere* deals with the frames and not the decision-making, but integration, solidarity, and the modes of incorporation include the need to account for the process of decision-making, which sits on the side of agency rather than structure. Which "decision is made" rests a lot on "a group choosing" a position between two or more alternatives based on knowledge about the alternatives and decisions that these alternatives represent—we will elaborate further details below. In complex civil societies, organizations and institutions are, necessarily, very inflexible. They produce path-dependencies and trajectories that are hard to circumvent and even harder to change. In some societies, bureaucracies have even erected the Weberian Iron Cage (modern day German academia, for example), while creating *anomia* between its cracks to a point which renders the category of society itself no longer applicable: Instead of the mythical "death of the state" we create post-societal nations.

Nonetheless, a more or less functioning civil society contains intermediate entities between organizations and institutions on one hand, and purposive (out/core) groups and individuals on the other. These entities can be described as civic networks that play a crucial (meso/intermediate level) role in decision-making processes; they depend largely on the existence of civics. We draw our definition of "civics" from the idea that citizenship is attained by education and maturation and not bestowed by birth. With the rights of a citizen come responsibilities, which must be understood and accepted and which basically include conceptions of justice, government and ethics. Without civics, the concept of citizenship is passive to the point of emptiness. Biological civics, as described by Foucault, Rabinow, and Rose, suggests that education must include the understanding of biotechnologies of the self and of biopower, as well as of the biological citizen in relation to the environment. The potentials of autonomy that biological citizenship supposedly offers, as its promoters claim, can only be realized if biological citizenship is complemented by biological civics. This is a relatively new aspect of the debate, which is gradually made explicit in part by the physical activity movement, the multiple intelligence movement, and mental health pluralism. Biocivic networks will be at the centre of future discourse on biological citizenship, biopower and governmentality.

Following Baldassarri and Diani (2007: 736), civic networks can be understood as "the web of collaborative ties and overlapping memberships between participatory organizations, formally independent of the state, acting on behalf of collective and public interests." This civic-networks-theorem manages to account for the interdependencies, polycentric structures, and variety of social bonds and transaction boundaries that effect actual decision-making on a social level while transcending individual, dyadic interactions. Based on empirical and historic evidence, the number of social bonds that civic networks uphold should, however, be smaller rather than larger (Baldassarri and Diani 2007: 774). The risk of fragmentation and factionalization rises with the number of social bonds and (simultaneous) transactions. Civic networks, then, become the facilitating media of incorporation on the level of action because they are the main staging area for complex decision-making in functioning civil societies. Furthermore, we need a better understanding of civic networking of this type. With regard to the aspect of biocivics that we will mention below, the embeddedness of civic actors and networks should happen at the community level. The social bonds they generate and maintain, while small in number, must then be necessarily mixed in nature and tied to the social/political/scientific importance of their transaction partners.

In civil societies with functioning civil spheres, narrative-empathic agents are not simply members of groups or representatives of organizations and institutions, but always are members of civic networks, just as all acting entities are narrative-empathic agents, participating in the civil sphere.

Solidarity as Empathy

Alexander (2007: 10) counts himself a member of a minority stream (including Durkheim and Parsons) in stressing the importance of *solidarity* for theorizing and for the *practice of democracy* in civil society. Secondly, he asserts that solidarity is value-neutral (in the Weberian meaning of the term) in reference to democratic society, for solidarity has both civic and noncivic forms. Finally, in considering solidarity on the macro-level, he uncovers the inner workings of solidarity and its links to outside social forces, revealing its dependence on a "sturdy" and systematic sociocultural foundation. Discourses about interrelationships, motives, and institutions are key articulators of this solidarity.

Alexander does, however, make the occasional—and arguably unavoidable—transgression into the micro-level. And on the micro-level, matters are not as clear: Even if we allowed for group actors to obtain agency, which is the very problem Pettit tries to solve, individual human actors are always agents; moreover, they remain as the primary agents. And it is this micro-level of actors that Alexander either refers to directly or fails to exclude in his discussion of solidarity and incorporation (Alexander 2006: 402/403). Calling on micro-level collective solidarity of individual human actors to account for any kind of action taken in the civil sphere requires a theory of agency that accounts for the property or ability of "taking someone's place imaginatively"—let's call this the *political imagination*. Therefore, this type of solidarity, "seeing you in others" or "taking their place," is an ability that is generally subsumed under the concept *empathy*. Empathy is the necessary property individual agents must possess to be able to act in solidarity. Unfortunately, like intersubjectivity, empathy is ambiguous and problematic.

Subsequently, it has been studied and theorized differently by neurologists, developmental psychologists, philosophers of mind, and others. In medicine, empathy has had a troubled career as a normative conceptualization because of the continuous pendulum swing between operational and phenomenological diagnostics and therapeutics in medicine and medical (including mental health) practice and education. This affected its career in other disciplines. However, with the rise of new theories to account for autism in neurology and the philosophy of mind, the problem of understanding others returned with full force, bringing empathy back onto the wider philosophical and sociological agenda. It brought along, however, ambiguities previously unaccounted for until Fritz Breithaupt suggested a way out of this dilemma. Recalling the different discourses of the first chapters, one can clearly see that from mathematical education to gamer culture and future children as organ donors, this richer concept of empathy that we are working with here does in fact apply to all of these fields (the empathy of mathematics, gamers, and unborn donors, etc.), because it is inherently practical. As such, it is capable of serving as the theoretical glue connecting practices that qualify as scholarly and techno-scientific.

Breithaupt manages to theorize empathy in exactly the kind of way that Alexander is forced to leave under-theorized for the agency aspect of solidarity. Breithaupt (unintentionally) complements Alexander's macro-level study with a micro-level or agentic concept that is highly useful, and together they lay out the road for new horizons of social theory and philosophy of science studies. To do that, we have to avoid the score of fallacies outlined in previous chapters and resolve the problem of intersubjectivity and solidarity discussed in this chapter (via a discourse theory of knowledge and decision-making) or effectively shove it out of the way by recasting it in a more realistic scenario than the contemplative and fictitious "Schuetzian enigma." Here we will see that it reveals the need for agentic saturation in Alexander (in order to go beyond Alexander) and, by contrasting it with Breithaupt's theory of empathy, we can see that, at least for a pragmatic social theory and philosophy of science studies, it can indeed be nicely shoved out of the way.

Signal to Noise: Narrative Empathy and the Cultures of Empathy

Empathy lies at the very heart of Breithaupt's concern. He hopes to provide a pragmatic approach to empathy that accounts for recent discussions in neurocognitive sciences while retaining practical and theoretical elements that do not simply "add" another position but try to identify empathy as an underlying capacity. Empathy, in his account, is narrative and not value-neutral: It is the very fundamental process of how we come to take a position at all; it is the practice of how to become un-neutral, so to speak.

Three Cultures

Breithaupt bridges the disciplinary gap from his field of comparative literature to cognitive science using empathy and narration. But since his discourse is deeply informed by neurocognition, psychology and the philosophy of mind, he provides a satisfactory theory of empathy and agency. In bringing together these disciplines, the three main accounts of empathy that Breithaupt discusses are: (a) mirror neurons and the idea of similarity, (b) theories of mind and simulation of other minds, and (c) the (coerced) adoption of perspective (Stockholm syndrome). He argues that all three phenomena point at three different and ideal-typical cultures of empathy; thus they cannot be reduced into one another. This irreducibility hints at some underlying empathic mechanism that accounts for all three types—a mechanism that he calls *narrative empathy* and which he declares to be "not centered on dyadic interaction," but on a minimally triadic constellation.

His observation begins with the fact that in our everyday lives we are constantly empathically challenged or, in his words, subjected to *empathic noise*: As social beings, we are constantly under pressure to "take up or consider perspectives" not our own, such as when participating in a conversation with a group of people, our own *empathic attention* constantly swings between interlocutors. Now, between our *neuronal activity and the expression of understanding, feeling with and for someone else, and compassion, lies the very (empathic) space of these cultures of empathy* that he seeks to explore (Breithaupt 2009: 9–10). The need for the *narrative element*, which is something that we as empathic agents add, is derived from the temporal nature of the interactions and the process that spans the empathic space.

While he does not say it explicitly, most of the other accounts for the phenomenon of empathy do not account properly for temporality and thus violate the primary criterion for good scientific accounts that Isabelle Stengers (2005a, 2005b, Stengers/Prigogine 1984, see also: Latour 2004) has described as an irreducible prerequisite for good science. In other words, through "temporalizing" empathic experience, we create a filter for empathic noise to be able to create the *illusion of an* insight *into another person, an illusion that is processed in the form of narration, the single and exceptional form of empathy allowed to be cut as a signal from the noise* (Breithaupt 2009: 12).

(a) Similarity and Mirror Neurons or Cultures of Similarity—Based on the study of neuronal structures known as mirror neurons, which function as *partial objects* (Deleuze) in this specific scientific discourse, and often based on the study of "feeling pain" and "understanding pain," neurophysiological researchers of cognition in their accounts presume that individuals assume that other individuals are (*somatotopically*) similar to them. However, Breithaupt shows that *empathy* cannot be accounted for by stipulations of this kind of similarity.

That we possess the ability for empathy and empathic agency is the reason why we as actors tend to hypostasize the idea of similarity to a point where it is rendered and *jargonized* into an explanatory pattern. In other words, it is because of empathy that *similarity* is turned into a phantasm, and the theory of mirror neurons as *partial objects* is but the contemporary translation of this kind of phantasmagoric hypostasis: Therefore, constructing the *Other* through similarity leads, Breithaupt shows, into infinite regress.

(b) Theory of Mind or Cultures of Construction—The problem of otherness that Breithaupt assigns to the *model of construction models* or *theory of mind* is that otherness must be (re)constructed in our mind. Consequently, the Other's knowledge and intentions have to be *simulated* with regard to the action perspective. However, the caveat is that such simulation has to be a one-point-construction: The number of differences between the actor/agent and the acting other cannot be too complex, but they must, ideally, be reduced to a singular aspect in the simulation. This leads to a problem for all construction models as accounts of empathy and for theories of mind as accounts for intersubjective interactions: We have to have deeper knowledge about the Other and his/her *inner being,* which brings about tautology and reification: *We learn to know what we already know of the Other* (Orig. in German: "*Wir lernen wissen, was wir vom anderen schon wissen.*" Breithaupt 2009: 82).

(c) The Power of Coercion and Reciprocity or the Culture of Coercion—Finally, using Stockholm syndrome as an illustration, coerced perspective-taking is the third possible alternative Breithaupt discusses. In this, he includes a discussion of both the force of coercion for reciprocal action and of gossip-theory (Dunbar 1998) as suggested in evolutionary biology. For all its flaws, this account delivers a new

idea, suggests Breithaupt, namely that empathy does not require two but (at least) three parties.

Alexander, Reich, Breithaupt

How do Breithaupt's *cultures of empathy* reconstitute on a micro-level account what Alexander's macro-account creates as *modes of incorporation* and how does it pertain to the resolution of "problems"?

Let us reiterate the three problems once again:

a. Reich poses the Schuetzian "enigma," how one can understand the other, leading to assumptions of similarity of understanding, referring to assimilation.
b. Kenny isolates from Durkheim the problem that solidarity, in order to account for a unified moral sphere, requires the stipulation of a "common humanity"—this refers to the idea that, *on par* with the particular out-group identity, a process of constructing hyphenation can take place.
c. Pettit proposes that group agency exists and related by supervenience to individual agency, meaning that the supposition of multiculturalism of groups overtakes the individual's out-group values on the action like the hostages in the Stockholm syndrome scenario.

Ad (a) For one, in the problem scenario of intersubjectivity or "Schuetz enigma," the fact is that for any solidary action we need to understand the person we show solidarity to: Reich, following Schuetz's "enigma," defines the intersubjectivity problem as twofold: (i) A person can understand the fellow-person, and (ii) the fellow person, the person who is to be understood, *must* cooperate. And even though Reich does mention Davidson, who provides ample formulation of this twofold problem, in his essay,[8] he hardly makes it clear enough that this is a problematic reversal of Davidson's use of the *principle of charity* (or, rather: it turns the principle into *transference*), in particular for Reich's discourse lacks a theory of cooperation.

A more general flaw that is revealed through Breithaupt's discourse, a flaw which also plays a role in Reich's reversed *principle of charity*, is that most of these (philosophical) theories that just *contemplate* and *construct* the "enigma"—we could say: without an account of decision-making and of the embeddedness of decision-making processes—construct it as a dyadic relationship. In Breithaupt's more sophisticated account, the fundamental and empathic situation is at the very least triadic.

Ad (b) The three modes of incorporation represent cultures of solidarity, but they themselves cannot account for solidarity; just as the cultures of empathy cannot account for empathy.

---

8 See also: Schützeichel 2009.

Now, what if the *modes* and the *cultures* are structurally similar? Would that not account for the postulate that the macro-aspect *solidarity* is at its very heart (a civil society with a functioning sphere?), related to the micro-aspect of narrative empathy?

(i) Assimilation and similarity[9]—The problem on the level of individual actors is that the practices in which actor A engages, in supposing actor B to be similar, will amount to a certain pressure for actor B to act in this expected way, otherwise there is reason for actor A to actively exclude actor B and to keep B excluded from future interaction. Lifting the first culture of empathy, *similarity*, up to the macro-level of assimilation constitutes the problematization of membership itself and is preconditioned by (1) the production of non-similarity on the micro-level and (2) the public–private distinction and the recognition of actor B of how s/he is expected to behave in public, even though s/he may not understand the underlying reason why s/he should do so. This means that s/he may act accordingly but not out of a sense of participation in the executed practice as being justified. Therein lies the source of many discontents: In an "assimilative" society, actors A and B assume that their respective other is similar to them in all *anthropological* respects and enforce a *primordial* pressure that only exists in public. In private, A and B remain unintelligible to one another without realizing it. For A, a member of the core group, private and public qualities converge; for B, as a member of the out-group, B's publicly displayed qualities converge with A but are not justified in the face of B's private qualities. At the same time, B assumes that A's private qualities would or should correspond to B's private qualities, therefore the divergence of public and private qualities is unintelligible but, through assimilation, is rendered discursive; B can participate (as a member) in civil discourse and express and pursue matters of public interest. Therefore, phenomena such as emotional contagion or false consensus effects (Breithaupt 2009: 25ff.; Hartfield et al. 1994, Marks/Miller 1987, Metzloff 2007) are distinctly related to the mere idea of the public.

(ii) Hyphenation and theory of mind[10]—The link between constructive empathy or theories of mind and the mode of hyphenation is easier to understand. Since hybrid discourses in the hyphenated scenario demand a limited bridging of the private/public gap, Breithaupt identifies the problem that in order to understand actor B empathically, Actor A must "learn to know what s/he already knows about B"—as in: *learning (about) difference*. Hyphenation rests on the idea that A possesses *a priori* knowledge about the private qualities of B and is able to simulate the private culture of B.

Jeffrey Alexander argues that, on the macro-level, instabilities remain, for the core group A does not accept B's primordial qualities as equal to its own. On the micro-level of empathy, this is accounted for by Breithaupt in the suggestion

---

9   The argument works similarly with Alexander's assimilation and Keen's in-group boundedness.

10   The argument works similarly with Alexander's hyphenation and Keen's ambassadorial mode.

that simulations are merely one-point simulations and divergences cannot be too complex or they would fail. In order for hyphenation to be an effective mode of incorporation, B or members of B need to understand something (have *a priori* minimal knowledge) of the private worlds of A as well, in order to not only react in public but to be able to understand the legitimation behind the practices. However, this understanding is not complete.

Ad (c) and (iii) Multiculturalism and Stockholm syndrome[11]

The relatedness of multiculturalism and the "Stockholm syndrome" culture of empathy is not as easy to see and also somewhat provocative. However, it allows for the suggestion that there are two types of multiculturalism possible, at least on the actor's level, that Alexander's account does not yet make explicit.

On the micro or interactional level, multiculturalism can be achieved either through a form of coercion or through autonomy.[12] Autonomy is understood by us in the Kantian definition, which is distinctly different from the (simple version of the) concept of freedom: here, autonomy means to find one's own justification for one's action and not acting as one pleases on a whim.

Multiculturalism is, in Alexander's account, a purifying procedure that affects the private sphere of core- and out-groups. Out-groups are made to purify their primordial qualities into a form that can connect with the core-group despite the differences and rendered into performances that are intended to *achieve* multicultural incorporation. This means that B, in *displaying* private qualities publicly, seeks out the explicit judgment of A, requiring A to accept the expression as a justified statement.

Alexander's model, due to its non-ideological nature, is rich and complex and he does not assume that multiculturalism is necessarily always a happy affair. Even in multiculturalism, exclusion is still possible, even necessary; and solidarity and integration are not identified with inclusion *per se*, but, rather, integration is constituted by inclusion and exclusion. Therefore, we can and should, distinguish between two modes of multiculturalism:

Multiculturalism Mode I refers to a mode of incorporation where B understands the *publication* and *purification* of its out-group values as a matter of survival. Insight comes by looking at the core-group—A—through the lens of the out-group's—B's—*relation with* A from an outsider's point of view, as the weaker observes the stronger, guessing not at what the A is thinking but guessing at what constitutes the relationship with reference to an outside world and assuming that A, too, reflects on the outside world and the relationship between A-B in reference to *this environment* rather than to B's intentions. Based on this judgmental situation, B (and A) can actively make the decision whether s/he (or as a group) "wants to keep playing with A (B) or not," in short, they can terminate

---

11   The argument works similarly with Alexander's multiculturalism and Keen's broadcast empathy.

12   On the semiotics of *coercion and* Weber—mind you, Alexander's theory of culture is a symbolic theory– see the work of Hans Bakker (1994, 2004, 2005, 2006).

the relationship or maintain it as a commonly shared discursive sphere. From this *holistic* point of view and accounting for the embeddedness of A and B in wider relations with the world and the environment, B can assess the consequences of cutting off communication with A or of being cut off by A. Thereby, B, in order to survive, needs to render its private qualities intelligible through performance to ensure long-term survival ("long-term maintenance of communication"). This requires understanding justifications for the difference of qualities. Purifying means "the *incorporation* of A's mode of justification by B." The goal, however, is not to change B, but to change A's attitude towards B (and others) using means available to A to facilitate the understanding of B's goals and values. This is a two way process, since in learning about A's means B cannot but try and understand what these means are made to do and thus consequently learns about A's goals and values.

But due to the nature of using this as a survival strategy, this means that B's integrity and survival is initially threatened; in a societal situation, where the core group is busy integrating out-groups, this looming threat is exclusion. Therefore, the process includes an element of coercion, albeit different from the one in the assimilation mode. The particularization of the universal that follows then leads to the mutual acceptance of diversity. This appreciation of diversity, however, is the result and not the cause of this process. There is, at least *ideal-typically*, a higher mode of multiculturalism—just as all three of the modes of incorporation and the cultures of empathy are *ideal-types*.

Multiculturalism Mode II is founded on narrative empathy, the possibility of a dialectical process between universalism and particularism, a concept of "truth as one *and* many," and moderate holism, because, ultimately, *pure solidarity* is *pure narrative empathy*.

Narrative empathy is the common denominator and basic form of agency upon which the three cultures of empathy rest (Breithaupt; Keen). If we assume that narrative empathy, despite its fundamental nature, can also ideal-typically occur in a pure form, then it may effect its own (sub-)mode of incorporation. However, rather than surpassing multiculturalism altogether, this mode can only improve upon multiculturalism. We would therefore suggest that, just like Kant's Enlightenment, we do not live in multi-cultural times, but in *multiculturalizing*[13] times: the pure form of solidarity that constitutes *multiculturalism mode II* is a regulative ideal we strive towards but will never attain, whereas *multiculturalism mode I* represents the form of multiculturalism that we can currently find empirically realized as the best possible scenario. The cultures of empathy are not only based on narrative empathy, then, but are also expressions of narrative empathy. However, they are also constraints of the pure form of narrative empathy.

---

13   The same goes for democratizing times. We constantly need to revitalize the pillars of democracy, which are never fully realized nor constitutive of our societies and polities: The day we assume it is, democracy is already lost.

As for the holistic perspective: Ultimately, achieving and performing culture rests on expressions and statements, which, as we have seen in the "Schuetz enigma," need to be understood (directly or mediated). While Reich has provided an insightful and rich overview of the problem from a phenomenological point of view, his account revealed some inherent weaknesses in the phenomenological point of view, most of all that its solution constructs its own problem tautologically: the problem of intersubjectivity is solved through intersubjectivity. Reich's own solution with reference to Sacks turns out to be a cul-de-sac because it brackets the heart of the problem and shifts responsibility for it to others.

The problem of how a person can understand fellow persons is, key to any theory of democracy, civil society, and solidarity such as Alexander's. Since Jeffrey Alexander has never subscribed to the kind of defeatism and silence that Reich prescribes, neither should we at this point. As a consequence, there is a need for both an agentic perspective on *Civil Sphere*, and a theory of understanding to account for key aspects of solidarity and integration. Now, for statements by an actor like A or B to be made or understood, discussions within philosophy in the past decades have made progress on various fronts but have also diversified within (Brandom 1998, Esfeld 2001, 2002, Pettit 1993).

Avoiding undue entanglement in these discussions, we can summarize a few key points (this outline follows Seel 2002):

a. Holism refers to the idea that any statement x relates to a set of statements that comprise a whole (X), and that understanding x means to know or understand (X).

b. It is in dispute what the consequences are, regarding to the expansion or definition of the whole (X).

c. Consequently, there are the following positions (Fodor/Lepore 1992, Seel 2002):

    i. Radical Holism suggests that (X) is infinite, but that it is still acceptable;

    ii. We have to repudiate any kind of holism, because is supposes (X) is infinite and that is not properly intelligible for a human actor (the position Fodor/Lepore themselves take);

    iii. Partial holism is possible for the actor, despite (X) being infinite;

    iv. Moderate holism suggests that (X) is indefinite, so there is a set of possible statements, which is considered to be a whole, but the set is never satisfied.

From the point of view of narrative empathy and *multiculturalism mode II*, we accept a moderate holism (following from Seel's account of Esfeld, and Pettit). From this, the condition that sets are holistic without an actual *whole* allows for iterations and inclusions in the set and renders statements potentially inherently communicable. The other options would render statements incommunicable or culturally problematic with regards to incorporation. This suggests that from a semantic point of view, "cultural incorporation" refers to the incorporation of these systematic sets of potential expressions as practices. The systematicity of expressed sets of statements can be considered to be *narrative* in nature.

Narrative Empathy

Real empathy is narrative empathy: only when embedded in narratives can the Other be intelligible to us. Triadic rather than dyadic in its relational nature, empathy arises from the conflict of (at least) two parties of Others. As a consequence, empathy cannot consist merely of "taking up and understanding someone else's perspective"; it is a conditional decision to take someone's position and thereby is decidedly *not free*.[14] This is because *taking a position* requires legitimation, and legitimations cannot be constructed, only called upon. In this respect, models of similarity, construction, or (en)forcible coercion are not inherently at fault—in actuality they represent the expressed *cultures of empathy*. However, not one of them is *the* sole account of empathy, for they are the actual results of the human capacity for *narrative* empathy, which, in turn, do not constitute an explanatory model, but merely the common denominator.

A particularly important issue that serves as the traditional point of convergence and contingency, and, subsequently, as the angle of discussion between (bio)medicine, neurocognitive research, the philosophy of mind and applied (bio)ethics, is the problem of pain. The connection of pain, (human) agency, and narrative presents the opportunity to reflect on the following relational condition: narrativity as a basic condition of human experience that exists embedded in and mediated by techno-scientific practices. This relational procedure of embedded mediation ranges from psychology, social ecology, and philosophy to architecture, medical imaging, and ICT (information and communication technology) design because "people live their lives as a narrative" (Sparacino 2002: 10).

Narrative can mean a number of things, of course. With an eye to the long history of science, political thought, philosophy, the *history of ideas*, Henrik Schaerfe reminds us that narrative space (the space that narratives both construct and inhabit, and that they occupy and furnish) entails "possible worlds," a trope that philosophers of logic from Feigl to Kripke and Goodman have inherited from the metaphysics of Leibniz, "who coined the term as an instrument for dealing with the theodicy aporia; the problem of *pain*" (Schaerfe 2002: 1, our emphasis).

Schaerfe's discussion of possible worlds and narrative spaces is full of interesting insights that expand the Leibnizian frame in his effort to combine different narratological theories. He argues that both narrative space and possible worlds resemble Douglas Hofstadter's recursive "stack" concept, flavored with the modal structure of narrative universes suggested by Marie-Laure Ryan (Ryan 1991, 1999, 2011, 2012, Schaerfe 2002). Schaerfe's ultimate goal is an

---

14   Unlike the Habermas/Honneth-style that (normatively) prescribes freedom.

exercise in formal ontology building: "an ontological description of the modal structure of possible worlds in narrative discourse" (ibid.).

Going beyond ontological questions, we are more interested in the fact that a "perceptual world" as a sequence of events "comes into place" through organization: How this organization is facilitated is problematic. The traditional point of view shared by diverse writers from Habermas to Bakhtin (even including some deconstruction) comes by way of "textual organisation," whether by logos or dialogos. Of course, they are correct insofar as there is a deeper root to this aspect that *also* leads to textual practices. This deeper root is, as Millar argues with reference to Bakhtin: "a beginning in Kant's categories of the mind organizing sensory experience to produce meaning" (Millar 2003).

While we agree with the above, we would add that this is not just true for dialogos/logos-based/textual practices, but also for the practices that are Real (Grosz 2011), textual and non-textual, discursive and non-discursive, dia-logical and dia-physical. And they are narrative and semantic, not because or when they are textual; textual practices are a special case of narrative, semantic practices. One even more important form is techno-scientific practices, which—as Mukerji (1998a,b, 2002, 2003, 2006, 2010), Carroll (1996, 2001, 2002, 2012), or Hengehold (2007) show—are inherently intertwined with state-formation, body concept, and the political imagination. Taking this further, the organization of possible worlds through techno-scientific practices is based on figured worlds. To some degree, we are enabled to say that possible worlds and figured worlds are co-extensive. Stingl's concept for the description of the affectual relations in medical imaging between doctors, patients, and patient life-courses, the concept of *narrative dialectics of techno-scientific seeing*, is more generally a case of semantic agency (Stingl on SAT in Stingl 2011a) that could be called *narrative dialectics of techno-scientific figuration* or *narrative dialectics of techno-scientific affect*.

Techno-scientific practices are semantic because they have meaning, and because they have meaning they are events that are sequential (Stingl), and the order of the sequence is narrative because narrative, in its minimalist sense, is a representation of at least two events with a temporal ordering between them. (Tversky 2004).[15] Tversky's account delivers a rationale for narrative *in the space*

---

15   Tversky's paragraph deserves to be quoted as a whole: "Life is happening all at once, all the time; everywhere, noise, light, smells from all directions, stimuli from without and within. Where am I, what is round me, where have I been, where am I going? What am I doing, what is happening around me. Our experience of coherence belies that chaos. Out of the stream of sensation, the mind craves objects into space, and actions in time, and configures objects into scenes and actions into events. If narrative is taken in its minimalist sense as a representation of at least two events with a temporal ordering between them [...], then maintaining awareness of space and time entails creating a minimalist narrative from the ubiquitous multimodal barrage if sensation" (2004: 380). A fuller, and yet complementary and not incommensurable, concept of narrative would include "[...] a

that omnipresently surrounds us (she appears averse to "environment"), that makes the events, opposite to the space itself, linear: like attention, they take things one after another (ibid.).

Together, this parsimonious conception of narrative and the notion of *figured* worlds suggest that the organization or sequencing of events can produce sets of rules. These sets can and will be acquired by actors based on how complicated the embedding of events is in "surrounding spaces" (physico-chemical, social, cultural) or types of environments. Learning occurs in interactions with the physico-chemical world in socialization and in enculturation. We could call these "*the* environment" as a cognitive whole due to the existence of these rule sets, but this requires caution (see also the discussion between Weiss and Restivo in Chapter 3). Hence, we arrive at a demand for semantic holism or social holism that accounts for and maintains the inherent "epistemic vagueness" that is required to enable learning, adaptation, and transformation in unforeseen future circumstances. It is suggested, again, that we require what Martin Seel described as a moderate holism (such as conceived by Esfeld, Petite, or Brandom).

Moreover, Tversky points out that the "[c]onsistency of perspective is also presumed to be necessary for construction of mental spatial framework in which to place each object or landmark" (ibid. 381). With regard to her "landmarks," should we not then be able to say that "landmarks" or "local signs" in medical narratives *are* the original basis for those symptoms that eventually appear in nosologies, that it is their reification and sedimentation on nosological maps that constitute illness narratives and diagnosis/care management pathways in acts of *Verdinglichung*, of objectivation or reification? It is, but this will only become possible to say in the form of a critique, instead of as mere criticism, when we introduce the notion of empathy as *narrative empathy.*

In order to create a genuinely pragmatic and applicable[16] concept of *narrative empathy,* we will have to accept the minimalist idea of what a narrative is. Plus, we have to emulate a concept of narrative that is pragmatic and practical: The narrative style, the mental framework or perspective that Tversky introduces is based on empirical research. Based on descriptions that people produced about their apartments or about city tours in which they participated, Tversky and her colleagues identified two distinct types of perspectives of narrative ordering plus a third, hybrid perspective (381ff.):

Route Perspective: Following a tour from one object succeeding another by linear change of viewpoint.
Survey Perspective: Description of landmarks from a stationary, externalized viewpoint.

---

narrative structure to have causal relations or narrative voice or demand narrative content to include character or emotion" (ibid.).

16   "Applied" is to be understood here in the way we would understand it in the notion of "applied ethics."

Gaze Perspectives: Hybrid in the fact that "like a route perspective, the spatial relations are relative to a viewer and like a survey perspective, the point of view is constant. In a gaze perspective, landmarks are described relative to each other from the viewer's stationary point of view in terms of the" (ibid.) viewer's spatial orientations (such as left, right, back, front).

Tversky (2004) compares the route perspective and the survey perspective: the former links at least two events in time, thus conforming to a broad sense of narrative, while the latter links declarations spatially rather than temporally. Additionally, even though space is static, a type of temporal order is imposed on stationary relations when it is described dynamically to a listener.[17]

We must analytically distinguish two very different levels here: *narrative* and *fiction*. This is important because in social phenomenology (cf. Alfred Schuetz) the concepts of narrative, fiction, and story are not clearly contrasted. By constrast, Tversky offers a sharply defined, empirically valid, and applicable concept of *narrative*. *Narrative* must be more minimalistic and fundamental than the concept of *fiction*.

In short, *fiction* is about beliefs whereas *narrative* is about practices. However, *fiction* is necessarily restricted to beliefs that can be expressed *textually*, which refers to both speech and scripture (spoken and written levels of language), but no more. Following in the footsteps of Foucault, Deleuze and Barthes (as well perhaps as Lacan, Irigaray, and Bourdieu), we could consider language to be more than *just* those two types of practices—speech and scripture. Generally speaking, our conceptualization of language refers to sets of expression that are at the same time structured and structuring.

This sets up a sequence of understandings about language where we recognize it as (a) a more comprehensive concept that (b) is intrinsic to Foucauldian and Bourdieuean notions that conflate language with spatiality and contrasts with a Deleuzean temporal comprehension that puts narratives and events in series, and that (c) can be reconciled with a moderate semantic holism that enfolds an understanding of semantics into a richer concept of language. As a result, those speaking of "text" refer to the limited idea of language that refers only to sets of oral or scriptural expression practices: In conclusion, all fictions are narratives but not all narratives are fictional.

From this, we can discuss a clearly defined concept of narrative: Narratives as sets of practices constitute meaning for actors as semantic, we distinguish between semantic and narrative levels therefore via the introduction of agency. The property of agency comes into play with semantics (and somatics) but not with narrative per se. Actors' actions become meaningful only at semantic levels where they are relational and performative because things that happen in universe between objects *just* happen, resp. *just* occur. We can then distinguish between

---

17   These perspectives should or narrative styles could therefore be reconceptualized as analog and digital perspectives as conceptualized by Ron Eglash (1999).

occasions—what just happens and occurs—and situations that involve actors and networks. Therefore, epistemic communities and their interactive context (such as the biomedical research or clinical context, or the mathematics classroom) involve actors and networks that need to be additionally qualified by making explicit (problematizing) the narratives in play (for example, patient narratives or illness narratives, or the master trope of mathematical rationality) that involve rich sets of textual *and* non-textual practices, and which can each be distinguished as different narrative spaces/sites and narrative styles with the inclusion of semantic and somatic spaces and styles.

If we look at the example of the experience of pain and the subjectivity of suffering, we can now say that while the *fact of pain*—a sensory perception—occurs inside an organismic (first order) nature, pain provides information in the form of signals. These signals needs to be ordered into data or event sequences that are *encoded, enacted, and embodied* as episodes of "suffering." In other words, suffering is a mode of *narrativization* or *event-sequencing* (see also Weiss's discussion of *mindscape* in Chapter 3).

On a performative level, this does involve imagination:[18] different actors are equipped with an imagination, which is both individually theirs and shared in common. This (political) imagination is, in Spinozistic or Deleuzean terms, a *univocity*. In the example of biomedical subjectivities,[19] because of the *univocity* of imagination, the politics and subjectivities of suffering are founded in different modes of *narrativization* that construct different illness narratives and patient narratives. This is a "politics in the third person." Esposito (2012) uses this concept of the "third person" to explain Foucault's and Deleuze's attention to the question of how the impersonal can be used to deconstruct both the dispositif of the person and the practice that changes existence. Foucault, like Kant, proceeds by way of a negative anthropology (Hengehold 2007) or negative instruction (O'Neill 1990). Deleuze proceeds differently. He begins by stating that "pluralism is monism" not just at the same time as, but *because of* temporality. He is, therefore, echoing Michael P. Lynch's insight that truth is "one *and* many." This statement demarcates biopolitical immanence, the consequential point of no return for any epistemic community whose practices are bounded by biomedical as well as by *anthropo-*, *bio-*, or *zoë-cological* concerns:

> For Deleuze, immanence is not generated dialectically by transcendence, as in Hegel, nor is it traversed by transcendence, as in the phenomenological or in the Heideggerian tradition. Immanence is nothing but the fold of being onto itself,

---

18 This, of course, leads to questions of the politics of suffering and the political imagination (see also: Hengehold 2007). It also shows how (see Weiss in Chapter 3 on Kirsh) epistemic and pragmatic acts are intertwined in a *narrative dialectic*.

19 Although this also works with mathematics, as Deleuze and Guattari illustrate in their discussion of *multiplicities* and Riemann (see also here, Chapter 1) in *Thousand Plateaus* (2004: 36f.)

its declension into becoming. This is what life is—it is always *a life*.[20] (Esposito 2012: 18)

As a consequence, semantic efforts concerning either life or truth usually involve finding ways of making narratives cohere. This is why *empathy* as *narrative empathy* is so important for any pragmatic, moderate holism of techno-scientific practice, governance, and material semiotics, aka *ScienceCraft*: Since we are dealing with political and social settings, when dealing with (techno-) scientific practitioners of any kind, a non-trivial group of actors and stake-holders we are looking at—*pace* actor-network theory—are human decision-makers. Making decisions, we argue (with Breithaupt), means first of all to make decisions "between different *narratives*" and, therefore, the situations we analyze in the (philosophy of) science studies involve what is called *narrative empathy* because the decisions are figurations or forms of integration between these epistemic communities of practitioners. They are figurations that must be effective, meaningful *and* sensible.[21] In other words, these modes of integration comprise our concept of solidarity, wherein solidarity is roughly defined as "an orientation towards others." In action, any solidarity that involves decision-making with regard to others, i.e. exclusion/inclusion, entails components of *empathy* that are spatial and temporal modes of narrative.

First, there is the construction of delimited reach or of space as suggested in Suzanne Keen and the three modes of *bounded, ambassadorial*, and *broadcast narrative empathy* (Keen 2011: 364f.). *Bounded* empathy focuses on in-group members, *ambassadorial* empathy targets specific others for a represented group, and *broadcast* strategic empathy calls to all potential recipients, usually regarding universal concerns.

Secondly, Fritz Breithaupt suggests that there exist modes or cultures of empathy that are based on the triad; this allows us to construct social interaction as a triadic relation as well, wherein decision-making by an agent manifests as choices between narratives.

From a social structural point of view, what matters is that Keen's approach allows for a spatial component for "integration-narrative empathy" that correlates with our interpretation of Foucault's discursive concept of language/space as heterotopologic. Meanwhile, Breithaupt's component allows for a discussion of institutionalist isomorphic structures (Meyer 2005, Schmidt 2012). At the same time, this includes (a) the institutional semiotic concepts of semantic and rational isotopia, and (b) a narrow concept of the narrative as "representation of at least two events with a temporal ordering between them." We also think that Deleuze's concepts of *attractors* and *temporalities* can be brought to bear within this theoretical frame. In this triadic account, an actor decides between two (or

---

20   Or as we would like to say: *aLife*.

21   On sense-making, enactment, and contingency, see: Karl Weick (1988; also: Eisenberg 2006); on meaning-making, see: Mieke Bal (1994, 2001)

more) opposing actor narratives, i.e. between two or more different options for the (temporal) ordering of events.

Narrative empathy is a contingent form of decision-making taking place in (minimally) triadic scenes; consequently, it has not just a *social*, but moreover, a *political* function. Empathy, according to Breithaupt's account, is not simply the ability to take the perspective and "feel" like some or any *Other* (not: *general Einfühlungsvermögen*), but a binding force and boundary-constituting process for socializing (*Vergesellschaftung*) and delimiting, even a segregationist force. Empathy is the process that con-figures inclusion and exclusion[22] for polity formation. Thus, being empathic does not equate with being moral. Quite the contrary, empathy, the very "practice of judgment," is the means for the generation of morals, and, therefore, empathy is "beyond good and evil."

In the practice of taking (or *voicing*) a position, empathy is also *exclusive*. Taking position for one Other means to take position against the other Other. To emphasize again, Breithaupt accounts for temporality, not just on the conceptual level, but also a second time on the practical level (and against the general trend of hyper-static, discipline-oriented [hypothesis-based] methods). The empathic agent can, at a later stage, switch and take the other Other's position. For analytic precision, we distinguish between temporal flexibility and temporal plasticity: temporal flexibility refers to our engagement with the other Other's position on a different subject/conflict while temporal plasticity refers to this engagement in a same subject/conflict.[23] Therefore, Breithaupt's *open*[24] conceptualization seems to us an adequate starting point for the study of the complex process of community formation (*Gemeinschaftsbildung*) through the employment of the boundary transaction functions of civic networks—far from any utopian ideologies or static discipline-oriented (quantitative) positivism.

Therefore, we feel confident in ascribing to Breithaupt a type of critical and structural realism that can serve as an action theoretical complement to Alexander's study, which can account for and help alleviate some discontents and faction-building in the sciences.

---

22   With a narrower focus on the neuro-sciences (mirror-neurons) but interesting conclusions such as a call for a "biosphere politics", see Jeremy Rifkin's *Empathic Civilization* (2009, 2010).

23   Given that we seek to reconstruct an interaction event, plasticity could be the focus of our interest. As for wider political and social implications, we might be interested in flexibility to account for a policy change with regard to group integrations that also suggest that positions can be embedded in systematic sets of beliefs. Therefore, flexibility is a necessary concept that accounts for changes of positions on a single issue related to wider systematic sets of beliefs; this marks the sociological group level necessary for Alexander's account.

24   We differentiate strictly between opening and closing accounts and conceptualizations, with closing being a distinct property of discipline-oriented science and opening being constitutive of problem-oriented research and scholarship.

Biomedical Progress, Social Inequality, Doctor-Patient and Narration:
Civic Networks and Medicine as a Civil Religion

In today's so-called digital, global knowledge, and information society, fraught with digital inequality and biological injustice, we have just begun to develop an understanding of (digital) biological citizenship.

A study presented at the 2010 Annual Meeting of the Association of Psychological Science in Boston, based on a screening of 14,000 students in the US, claimed that students' abilities to empathize were rapidly declining.[25] According to the authors of the paper, empathy levels among students have dropped by 40 percent since the late 1970s, corroborating a "strong feeling" by those of us in the academy about declining empathy levels among our students; this feeling is supported by other indicators as well. We conjecture that recent social and political developments around the economic recession/depression that unfolded in 2007–2008 contributed to these reduced empathy levels; more profoundly, the severe cuts that are affecting schools and universities have tied our hands in fighting this downward trend.

Even if the type of empathy discussed in this study refers only to the first or second culture of empathy that Breithaupt describes, there could be larger ramifications. It is a logical conclusion that the dynamic properties of the underlying concepts of solidarity and narrative empathy are not merely "affected," but that their acquisition and practice by recent student generations is impoverished. Like Kant's notion of judgment, perhaps because of a certain epistemological heritage, *narrative empathy* may need constant practice and exercise.

Not only do we need further study into the decline of empathy in our societies, but we also must enrich such studies with a more complex understanding of empathy and its ties to solidarity (a) to get to the bottom of the problem (which any under-theorized concept of empathy or agency will not help us do), (b) to weigh the consequences of this development for the future of civil society, and (c) to develop effective civic counter-programs to promote solidarity, autonomy, and democracy.

In this brief concluding section, we are going to illustrate this issue with regard to biological citizenship and practice in medicine. Although applied to this specific topic, we urge readers to recognize the deep theoretical mechanisms connecting to neighboring discourses and techno-scientific practices such as the discussion of mathematics and civilization, the mathematics classroom, the mind-body and the social, as well as nature-nurture debates, gaming and gender issues, and the present and future of unborn and even "sibling savior" donors (see the preceding chapters 1–5).

The discourses of biomedicine and civics are excellent grounds for illustration and also for intervention for several reasons. It is readily apparent that biomedical progress, in recent decades, has had the most profound effect on our understandings

---

25   Available online: http://www.ns.umich.edu/htdocs/releases/story.php?id = 7724.

of citizenship, democracy, and ethics beyond other developments, for it interacts with all other domains of social and scientific progress and their discontents, such as (global) environmental issues, the effects of *digital divide* inequality and injustice, and the profound ways that religious belief systems of many people have been challenged (from the resurgence of creationism and reproductive rights issues to new end of life questions). The biomedical component of the global discourse and within the epistemic imperialism of the Western colonial matrix of power has irrevocably shaped the political imagination.[26]

The progress in biomedicine, consequently, is intertwined with our lives as citizens in democratic countries. If the decline in empathy affects students in general, it also affects students of medicine. But isn't the folk concept of empathy—the idea that a medical practitioner *understands* not just illness but also (and most importantly) his/her patient's feelings and life—the most basic and most important skill needed for valid diagnosis and adequate therapy? And is it not the adequacy of a therapeutic regime that will eventually enable better care, establish better health, and, as a consequence, also reduce costs—which is known as the Triple Aim Agenda in modern health care?

The discussion surrounding different diagnostic strategies, the training of doctors, concepts of health, illness, and care has been as heterogeneous as it has sometimes been ideologically ferocious. The main division lies between operational and phenomenological diagnostics, with most of the other discussions theoretically and historically derived from this crucial opposition. This can also help differentiate between forms of social diagnosis and strategies for civil repair because both are cases of knowledge-based decision-making practices, and thus the basic mechanisms are the same. Phenomenological diagnostics, which has recently been considered as good as dead (Andreasen 2007), frames the human being as a biopsychosocial organism embedded in and interacting with different types of environments.

For a medical practitioner in the operational (Kraepelin) model, it is necessary to memorize (by ro[u]te) and follow a catalog of symptoms and predetermined therapy trajectories aligned with an existing health care management bureaucracy that aims to reintegrate the patient to a state of "normal functioning." In the second model, the medical practitioner instead attunes to the narrative nature of the practitioner-patient situation and seeks to understand (empathically) the patient's care-requirements. The goal is to integrate therapy with a patient's life-course in order to bring the individual life-course and life-style into consensus so the patient can have a chance at a happy, healthy, and successful life in accordance with his/her individual potentials. In his critique of the Kraepelin model, Karl Jaspers sketched the second model as an alternative *ideal-type* and declared that actual medical practice would realistically (have to) be situated between both ideal-types. Additionally, Jaspers—sometimes called Kant's last and probably only true

---

26   That the epistemic imperialism of the trope of biology would reach even into the study of International Relations (IR) has been effectively predicted by Duncan Bell (2006).

disciple by Hans Saner—was a proponent of a philosophy of democracy that we find expressed in the work of Hannah Arendt, Susan Neiman and Onora O'Neill and that is derived from the same ideas that informed his reflections on clinical practice and that strongly suggests an active concept of civics.

Jaspers' idea of *clinical practical reasoning* (CPR) is a version of narrative empathy that is actually practiced (by some really good care and healing practitioners). Secondly, a growing number of authors have begun to stress the need to reform clinical practice by enriching clinical decision-making with more in-depth (narrative) knowledge about individual patients and a more critical stance towards mere (material) technological innovations. At the same time, the decline in empathy among recent and current students seems to comprise a counter-development to these positive notions.[27] Because of the central role of biomedicine in contemporary civil society, the lack of empathic skills in the primary agents of biomedicine could be both an indicator and cause of an ongoing disintegration of civil society or of a new and as yet undefined emergence of the "dark side of democracy." Even Nikolas Rose (2006), for all his optimism, accounts for darker scenarios such as the "return of euthanasia through the back-door," to name one example. This is exactly why reforms in the field of health care and deliberations on progress in biomedical science and higher education in the field of medicine studies span one of the crucial discursive spheres where the future of modern civil society will be decided. They also involve a great many different types of narrative.

Kevin Dew (2007: 110) has determined that Durkheim's call for a new religion—which Bellah and others have transcribed as a *civic religion*—has an institutional candidate in medicine within Durkheim's framework. If based on the resounding calls for a "cult of humanity," Dew argues, the core idea of Public Health would fulfill that role since it connects the reduction of health inequalities with the reduction of social inequalities: because public health integrates basic functions like eating, drinking, and moving into a system of meaning[28] that contributes to a collective good, the resulting "organic solidarity" can form the basis for this civic religion.

This is, of course, a kind of public health that does not restrict itself to demographic registration, bureaucratic administration of budgets, and pre-existing care-pathways, but makes good in practice what has up to now only been a "rhetorical focus on equity, empowerment and community mobilization in healthy public policy" (Dew 2007: 110). Two questions follow:

---

27   Again, with regard to the example of savior siblings, participant decision-makers cannot let themselves be hijacked by a mono-culture of empathy. Doctors, researchers, and law-makers involved must not only create their discourse bounded by either "mirroring sympathy" nor assimilate the culture of the reduction of ethics to law. In between the narratives, they must be modest cyborgs who can connect with different narratives and decide between them.

28   Which is called dietetics in Kant, religion in William James, askesis in Foucault, somaesthetics in Richard Shusterman.

    a. Does democracy have a somatic dimension?
    b. Can this be sociologically and politically operationalized beyond theory alone?

As for the first question, Dewey's work is cited and revered by both Jeffrey Alexander and Elizabeth Anderson (2006). While not primarily educated in clinical psychophysiological laboratory culture as with first generation pragmatists like William James or Edmund Burke Delabarre, Dewey, when fleshing out his ideas on the public and democracy in his middle and late phase, was engaged in the public health debate, particularly in promoting the Alexander technique (referring to F. Matthias Alexander). Today, this plays an important part in the idea of somaesthetics, as discussed by Richard Shusterman, and thus answers (a) in the positive.

Regarding the second question (b), Peter Wilkin (2009), in a recent and rare exploration of just such an issue, has indeed demonstrated that such an approach is not just possible, but highly productive. Addressing the problem of individual good living and the good society, he showed how social policy, biomedical research, health care costs, loss of economic profit and individual health are tied to the industrial production of chairs, the rise in the incidence of lower back pain (LBP), and the economic and health-related effects thereof. From this, Wilkin illustrates how a scientifically informed, empathic and democratic strategy can lead to better living and a better society (and hopefully, better chairs).

This suggests that for the (techno-scientific) practice of public health, *narrative empathy* serves well as an ideal type. And in light of solidarity and civil society, the cultures of health, expressed in certain life-styles, must be incorporated in the same fashion as ethnic cultures that participate in the civil sphere. As the citizen has become biological, so too has the civil sphere. Thus, to confront contemporary dangers and discontents that threaten to bring the democratic heart of our civil society to a stop, a *public healthy diet* may be called for before we must resort to actual CPR.

Integrations

In our complementary, rather than critical, appraisal of Jeffrey Alexander's *Civil Sphere*, it represents the re-introduction of a pragmatist voice into the discussion on how the practice of solidarity can facilitate social integration in contemporary democratic societies—an approach that navigates cautiously between the ideological Scylla and Charybdis: neo-liberalism and neo-Marxism.

His discourse might then enable a new and improved pragmatic social theory, but only if his macro-theoretical discourse is supplemented with a) an enriched theory of agency and, in order to account for the specifically modern condition of information society, b) a sufficient theory of knowledge and truth economies. We have attempted to provide an account for (a), the micro-level problem of agency, based on Fritz Breithaupt's concept of narrative empathy, by leveling three problematic discourses pertaining to the conceptualization of solidarity and understanding. These three problems have (re)emerged in recent theoretical essays on some classical issues: the interest in Durkheim's cult of humanity (separately addressed by Robert W. Kenny and Kevin Dew), Wendelin Reich's illustration of the intersubjectivity problem known as "Schuetz's enigma," and Philip Pettit's elaboration of the existence of group agency. Each of these problems corresponds with one of Alexander's three modes of incorporation and one of Breithaupt's three cultures of empathy, as well as with Suzanne Keen's concepts. On the macro-level, the modes are derived from solidarity; on the micro-level, the cultures are derived from narrative-based positioning. Taken together, these two modalities may help resolve the issues rising from the three classic problems. Moreover, in the final section of this essay, we argue that due to its biopolitical nature, contemporary democracy requires a biomedically informed concept of civics *that the Civil Sphere* alone might not be able to provide. We recognize that Alexander's concepts of understanding and solidarity in the *Civil Sphere* are uncharacteristically undertheorized.

We want to go further and postulate that the combined power of Alexander's theory of civil society, a theory of social agency, and a theory of knowledge economies could give rise to a new pragmatic and efficacious social theory of the integration of techno-scientific practice. This, more complex and pragmatic social theory accounts for empathy as the basic modality of agency in democratic societies—combined with a rich and complex theory of knowledge and its economies—and might attain the theoretically saturated levels necessary for our age of biological citizenship. Such a social theory would be mature and complex enough to account for the problems that arise with the age of biological citizenship, digital divide, the global epistemic culture and the new forms of *glocal* legislation. It could account for the contemporary emergence of *biopolities* because it would present as a sophisticated theory of *postcolonial techno-scientific polities* at large.

Summary and Conclusion

To provide reasons for moving towards an agentic saturation of Alexander's civil sphere, we discussed three problems that (re)emerged in recent sociological and philosophical literature: the Durkheimian moral problem of solidarity and humanity, the "Schuetzian enigma" of mutual understanding (intersubjectivity), and the problem of group agency. We then moved on to argue that there is a structural similarity between these three problems, Alexander's three modes of incorporation, and Fritz Breithaupt's three cultures of empathy. Breithaupt's search for a fundamental form of empathy allows, in our account, for an agentic solution to discontents posed by the three problems for Alexander's theory in practice.

In the fourth part, we argued that Durkheim's original fascination with medicine (which was largely shared among the majority of nineteenth-century scholars) offers a main stage within contemporary society and the field of biopolitics for discussing and operationalizing Alexander's civil society and Breithaupt's narrative empathy. The practical implications from this can be felt in the education and redemption of the "medical practitioner" vocation since it is they who hold key intermediate positions within *biopolities*.

The role of the third component of a *new pragmatic social theory*, knowledge, has not been discussed in this chapter, but here we offer one key aspect: What is important to understand about knowledge and knowledge economies for this kind of social theory and the creation of a positive culture of *biocivics* is that the role of state and government agencies lies not in the control of knowledge, but in preventing—through regulatory mechanisms—the monopolization of access to information by any one entity. A *bio-civil society*'s government's task is to keep the channels and capillaries, through which information flows and knowledge is generated, open and flowing—*pace* the *digital divide*. This can happen only if the society itself has a functioning *civil sphere* cognizant of the question of *biopolitics*. In introducing the concept of the *civil sphere*, Jeffrey Alexander has helped us to better understand what a civil society can be and what can be achieved through civil society on the macro-level. Adding *narrative empathy* and the *cultures of empathy* into the mix, we can understand how civics and civil action work on the micro level. In other words, we do not yet live in civilized times, but in an age of making the civil sphere come to life.

Here, we pick up our argument from Chapter 4: Neither the civil sphere or the techno-scientific sphere can build Truth (as one and many) upon a single voice or perspective. As polities—a figured world and social ecology spatially and temporally governed by techno-scientific practices and imagined in their narratives—civil societies and epistemic communities of science and scholarship would be doomed if they were monocultural, just as biological monocultures are doomed to failure and collapse. The biopolitical citizen, the pragmatic scientist, the multi-disciplinary scholar: they all must learn to see through the multi-faceted prism of the modest cyborg. Recall, the idea is not only to bring a voice into the cathedrals (of science), but also to invite Others to sing. Neither assimilation nor unity through blind purification, *univocity, the pluralism that is monism,* can be harmonious and dynamically dissonant, celebrating diversity and transformation.

Epistemic communities, such that exist among the practitioners of science and scholarship, should therefore be wary of monocultures of the mind. That doesn't mean that they need to accept that "anything goes" without restriction. Nor does it mean that multi-culturalism requires that a community sacrifice its identity. It means that these communities should share in solidarity and aim at integration in the wider civil sphere because their (techno-scientific) practices help con-figure the political imagination of those who populate the civic sphere as human decision-makers. Creating this kind of solidarity requires us to make explicit the mechanisms and consequences of integration and decision-making; we have tried to do that here.

For epistemic communities of science and scholarship that are con-figured in techno-scientific practices and governance, the modes of integration represent how knowledge and knowledges become integrated. It is easy to see that the *assimilation* of knowledge requires that inclusion must occur by an act of purification that destroys the original and primordial qualities. An excellent example is the assimilation of Chinese Medical Practices into Western Biomedicine and Alternative Medicine. When they cannot be dismissed easily by the Western biomedical community, Chinese medical practices in Europe or America are often investigated while stripped of their social-ecological context and individual patient modeling. Fixated on *the* "active ingredient," investigators often render Chinese herbal medicine irrelevant and impotent (and consequently, subject to dismissal by skeptics). This is a common pitfall because fundamental ideas, such as the notion of *Qi*,[29] are not easily translated into Western vocabularies or regimes. Meanwhile, alternative medicine enthusiasts often omit the historicity of Chinese medicine—its collective *empirical memory*[30] that spans thousands of years while it has, at the same time(!), existed only since after the *Cultural Revolution in China*. Many practices, such as acupuncture or Qi Gong, are taught by hyphenated-practitioners, Western physicians, or healing practitioners who have taken only a few courses. While they claim that these practices work, they do not understand how or why, nor do they care. Because of this, these practitioners can never fully replicate the expertise and efficacy of the original practice; they still structure the practice according to the Western colonial matrix, which includes the "black-box" or residual category of the placebo-effect (Cohen 2002). A simple *multiculturalism mode I* would point to the jealous co-existence in parallel with different communities. This can be found in cases where patients participate in different medical regimes but do not dare tell either practitioner of the other's involvement. Consider the case of a hemorrhoid patient who simultaneously applies a prescribed pharmaceutical based creme from a Western doctor while adding essential oils provided by an aromatherapist. The interaction of these different substances, which have very different chemico-physical properties, can lead to unpredictable, possibly damaging effects.

The *cultures of empathy* suggest a similar account. A culture that assumes in the *Other* general similarity will run into severe problems. Because medical practice requires benchmarks, such as the use of anesthetics before surgery, benchmarks are defined as standards. But if the standard is produced based on the idea that the *generalized Other* will be biologically identical, then the danger is clear: A small Asian woman is physiologically not the same as a tall European white male or a South-American transsexual and will not respond identically to the same regimen of pharmaceuticals nor present the same physical readings in response

---

29   The most pragmatic translation that we have found so far is the concept of *agency*, that is used in recent discourses on *New* and *Vital Materialism* (Bennett 2010, Coole/Frost 2010, DeLanda 2002)

30   By this we mean the generation-spanning non-verbal inheritance of practices, inclusive of replication errors.

to similar interventions. While the basic idea of moderate holism implies that we have to assume that an *Other* has a biological equipment not unlike our own, the convention is not that it is similar (to the point of identical) but that it is not "bizarrely different" in its disposition. To have a theory of mind or a simulation of the *Other* means that we have to have knowledge of them. This should alert us to the question of the social ecology (see Chapter 3).

The matter of coercion is found in the Western Colonial matrix of power, in particular the promises that it brings. As a hostage is "coerced" into siding with the hostage-taker, the short-term relief from painful symptoms may seem to justify a unilateral implementation of Western medical practice. But if that practice does not help deal with an underlying (social) context that presents the actual cause, where then is its advantage over a traditional method that addresses either or both? Looking to another example: while modern methods of agriculture promise a more profitable yield (also a Western, modernist way of measuring success), traditional methods of farming may be more sustainable (assuming access to the knowledge-practices). The issue is that there is no one True way of seeing and of practicing; no one culture of con-figuring the world is *the* one. The issue is, precisely, that above the information and data, knowledges are made that are not isolated from other fields of practice. It is their integration that counts: how decisions are made and how they connect with one another in the most pragmatically effective way. One must see the *Other,* as a techno-scientific practitioner, situated within his/her context. This can take different forms: a route (perspective) where one is living linearly through the problem, a survey point of view that describes the issue symptomatically, or a gaze that reflects the points of view.

In communicating with others we need to carve out a space for multiculturalism (mode II) so that through the integration of heterogeneity (e.g., the heterogeneity of theories of mathematical education in the *semiosphere* of the classroom, Lotman 1990, 2005) individuals can generate "uneven multi-cultural space[s] of meaning-making processes and understandings" (Radford 2008: 318).

How one spatially constructs this narrative realm is important as well—specifically, how one builds the boundaries of the cooperative-communicative aspect of practices. Often, techno-scientific practitioners remain in a bounded group orientation where they want to interact only with like-minded collaborators; this even carries over into material semiotic practices. Even in object-interventionist techno-scientific practices, we often engage with (epistemic) objects[31] only in the bounded form of narrative empathy. The proverbial case goes a bit like this: "Presented with an fMRI image and told to look for an aberration, one specialist for this kind of disease will see this, another that."

We do not want to entertain polemics here; we use this stereotypical example for illustration. However, we also emphasize that there is enough empirical

---

31   Keen does not restrict her concept to people, but includes "honored objects," which we translate to mean "epistemic objects."

evidence to support our argument that object practices and visual practices are severely constrained in techno-science, and that a different object-relation is just as necessary as a different person-relation. Reflecting on which narrative empathic orientation towards others and objects is currently being practiced and which (bounded, ambassadorial, or broadcast) would prove effective is a viable strategy to enable integrative processes between scientific communities.

Kusanagi Motoko, the dis(/re)embodied protagonist of the Japanese animated cyberpunk classic, *Ghost in the Shell: Stand Alone Complex* (Kamiyama 2002), noted that "(t)he mark of the immature man is that he wants to die nobly for a cause while the mark of the mature man is that he wants to live humbly for one." We need the services of a modest cyborg knowledge-practice rather than the sacrifice of a rigid bastion of purity: the political imagination of techno-science should continue to be scientific, but it should begin being solidary as well.[32]

---

32 *Postscript comment by Restivo*: This chapter, like all that goes before is our collective effort. It is more than any other chapter, however, a white paper on the future of social theory that reflects unique contributions within Alex Stingl's emerging paradigm for a nü-pragmatism.

Conclusion:

Harder Better Faster Stronger: An Epilogue in the Mask of Four Human Problems

Alexander I. Stingl and Sabrina M. Weiss

Introduction

Madison Square Garden, October 20, 2010: The French band *Phoenix* celebrated their newly established fame and success in the United States in front of a large audience. During one of their final songs for the evening, their hit single *If I Ever Feel Better,* they abruptly stopped, leaving their fans confused in the dark with only a machine hum as background noise. Suddenly, the returning lights revealed two figures wearing silver suits and robot masks. In electronically altered voices they sing-songed the audience into a frenzy:

> Work it—Make it—Do it—Makes us
> Harder—Better—Faster—Stronger
> Work it harder, make it better, do it faster, make it stronger

The two robots were the French musical and performance duo *Daft Punk,* who flippantly took their name from a scathing review of one of their earlier projects where their work was described as "a daft punky thrash" (Jennings 1993). For more than a decade now, both members of the band have not appeared publicly, whether in person or through media, without wearing a mask or being digitally obscured. It is no accident that we have chosen to borrow the title of their song for our closing chapter, in which we both reference and pay tribute to Marcel Mauss, Michel Foucault, and Donna Haraway, three names of scholars whom we might be willing accept as a *chiffre* for each of our names in this endeavor: Restivo, Stingl, and Weiss.

We are not the first to look to these cyborg performers for provocations beyond the artistic. Chad Parkhill, in his critique of transhumanism (2009), describes how *Daft Punk* has used their various works as "occasion[s] to meditate on the relationship between technology and the human" in a variety of ways. Their movie, *Electroma* (2006), is specifically explored as an existentialist critique of a central thesis of transhumanism—that death is a harm. Parkhill places *Daft Punk* in conversation with Sartre and Bostrom to reveal the circular nature of these critiques:

If we are to believe both Bostrom and Sartre when they claim that their doctrines are extensions of humanism, then we must recognize that an existentialist critique of transhumanism will only return us to the humanism that underpins them both; had they succeeded in becoming human, Daft Punk's robots may have found themselves dreaming of becoming robots once more. (Parkhill 2009)

Although it would benefit from diffraction through Haraway's conceptualization of the cyborg, this analysis illustrates the spiral movement reiterated in our concluding chapter as it was in the entirety of this book: Each time we gaze beyond the boundary, we find ourselves staring back at ourselves as we have been before.

This isn't so different from Foucault's effort to make himself heard above *La Belle Noiseuse* (Serres 1995) caused by his *work/Werk/oeuvre*, which was so loud that he was afraid that his audience couldn't properly hear (and understand) him. Though his words had created a space that was full of echoes and resonances that drowned his voice, he was even more afraid that he could not hear himself. So in 1980, he agreed to an interview with *Le Monde*, which was published anonymously under the title "The masked philosopher" (*Le philosophe masqué*). In it, Foucault expressed his hope for a more creative, productive form of critique:

> In our societies, characters dominate our perceptions. Our attention tends to be arrested by the activities of faces that come and go, emerge and disappear.
> [...]
> I have an unfortunate habit. When people speak about this or that, I try to imagine what the result would be if translated into reality.
> [...]
> I can't help but dream about a kind of criticism that would try not to judge but to bring an oeuvre, a book, a sentence, an idea to life; it would light fires, watch the grass grow, listen to the wind, and catch the sea foam in the breeze and scatter it. It would multiply not judgments but signs of existence; it would summon them, drag them from their sleep. Perhaps it would invent them sometimes—all the better. All the better. (Foucault 1997: 321ff.)

In masking himself, in hoping to redirect readers' attention to the meaning of his words, he became a cyborg, the mask his extension. Only in "lowering" his voice, in becoming cyborg, could he be heard at all, or so he thought. Scholarship must mean to *become* cyborg: Asked if he is "in fact for a society of scholars *[societe savante]*?," he answers (in a description suggestive of Haraway), "I'm saying that people must be constantly able to plug into culture and in as many ways as possible" (Foucault 1997: 327). That he both lowers his own voice and calls on others to lower their voices—a "year without names"—shows that Foucault was not just a cyborg, but a *modest cyborg*.

Inspired by these artists in robot masks—or philosophers in musician masks?—and our philosopher behind a mask of anonymity, we "modestly" ask ourselves what happens when the mask is taken off to both unveil and become the direct object of our gaze: Is the robot staring back at us, and is the robot—in returning its Otherly gaze—the actual person? What does it mean if the mask looks back at us and speaks back at us, unveiled yet modest? Is this, perhaps, the ultimate play of the mask—that it veils itself in personhood?

Although the mask and the person are two concepts that are deeply intertwined, it is important to avoid the temptation to conflate the two mindlessly. Giorgio Agamben, for example, leverages the etymological relation of *persona* to *mask* to define the *dignitas*—an image or mask—as a prerequisite for a human to participate in the legal and political spheres. This type of analysis, also shared by Roberto Esposito, reduces the mask and persona concepts to technics ("a device, dispositif, or apparatus") that allow juridical governance to be related to life (Parsley 2010). Marcel Mauss, by contrast, brings more richness to this discussion by retracing the history of the concepts *mask* and *persona* (rather than just the etymology). In concluding his essay on the notion of *person* as a category of mind, he warns of an overeager conflation of the terms while at the same emphasizing the importance of noticing the interrelated dynamics of their relationship:

> From a simple masquerade to the mask, from a 'role' (*personnage*) to a 'person'"(*personne*), to a name to an individual; from the latter to a being possessing metaphysical and moral value; from a moral consciousness to a sacred being; from the latter to a fundamental form of thought and action—the course is accomplished. (Mauss 1938/1985: 22)

He does not have time to explore the relations between the concepts *persona*, *imago*, and *simulacra*, which he reconvenes from Roman legal speech to Lucretius and Spinoza. He sets up the persona against the dualistic (or as we flag them, "binary") constructions of the world and the human condition. We cannot emphasize enough the importance of both Lucretius and Spinoza, not just for Mauss, but also for the development of Western thought and civilization. Moreover, we cannot ignore their inherent potentials for *epistemic disobedience* from inside the Western colonial matrix, nor can we ignore the foundations groundings (*arché*) that biology receives in these terms. For example, Stephen Greenblatt has revealed the connection between Lucretius's poem, *De Rerum Natura,* and modernity (2012) while students of *vital materialism* were already familiar with the importance of Lucretius and Spinoza for modern thought in general and, in particular, for Henri Bergson and Gilles Deleuze (see among others: Bennett 2010, Grosz 2011, Bryant 2012). And Lucretius's role in eighteenth-century German thought has been accounted for by Nisbet (1986: 99–100):

> Along with Spinoza, Lucretius provided the eighteenth century with one of its main models for a rigorously naturalistic explanation of all reality, and the radical Enlightenment with one of its weapons against teleological and Providential views of nature and human history. Lucretius's arguments against religion,

immortality and the fear of death[1] are continually cited by the *philosophes*. (Nisbet 1986: 99/100)

In the co-evolution of biology and anthropology (which did not yet have their names), psychology and philosophy in the eighteenth and early nineteenth centuries from Albrecht von Haller to Kant, hardly any notable scholar could have missed the importance of Lucretius's language. Spinoza's work has been given similar recognition through works by scholars such as Jonathan Israel (2002, 2011), whose account of the history of Enlightenment and Modernity shifts the center away from standard accounts to Spinoza.

The *imago*[2] itself is a central conceptual device to describe the final stage of insect metamorphosis, which any historian of biology cannot but associate with the name of Goethe (cf. Richards 2002). It is noteworthy that Mauss' interest in the relations between the individual (part) and its society (whole) was guided by Pierre Janet's and Durkheim's notions of the collective, a notion that also appears in Carl Jung's psychology in relation with the concept of the *imago*. What connects Mauss to Jung is an interest in the development of the self, the notions of "I" and "Me" (*Je/Moi*). For Mauss, the "techniques of the self" cannot be obtained without the environment (as a social ecology), nor without the (techniques of the) body. The human being is conceived, in Mauss' work, as a total human being, the individual conscious mind, a small layer. Though small, the mind *is* the intermediary between physiology and the social in this line of inquiry. It is also an "adventure of an idea," to paraphrase Whitehead; it is a project that is risky because it can never be fully realized. The total human being (*l'homme total*) is not (an) absolute; attempts to embrace this *totalité* and *integrité*, as in Mauss' scientific venture (forged in Durkheim's shadow), are heroic but doomed to failure (Hart 2007: 7f.). However, Mauss' admonition to approach this issue in the style of the humanities, rather than through scientific abstraction, holds promise.

Both the notion of *totality* and of *integrity* are not to be understood in a Hegelian sense. We consider them to be closer to Simmel's idea of *totality* and a version of Durkheimian integration, which we have elsewhere called the *anthropos*[3] (Stingl, forthcoming, 2011a, 2011b, 2012n, Stingl/Weiss 2013). But

---

1   Parkhill (2009) argues that most scholars favoring transhumanism, also see "death as a harm" and search "for immortality." Any genuine posthumanist vision, however, such as the vision of a critique of *Daft Punk*'s humanism through Haraway's cyborg would be free of such consideration, in, what we would like to call a gesture of transcending of the boundaries in the fashion of Bataille.

2   For a more detailed elaboration of connections between *imago* and notions of phantasm, see the works of Roland Barthes, Georges Bataille, and Walter Benjamin.

3   Although Stingl's *anthropos* concept (forthcoming, 2011a,b) was originally derived from Rabinow, we intend to incorporate, in future iterations, ideas from Walter Mignolo, who, along with Nishitani and Naoki Sakani, uses *anthropos* in relation to the concept of human being as determined by the Western colonial matrix, whereas *humanitas* is used as a more comprehensive concept.

Anthropos is not a constitutive (in the dualistic and Cartesian sense) or an absolute (in the Hegelian sense). Using Simmel's understanding of *totality*, we claim that the human condition as vital is composed of and comprehended by fragments that require completion by other means. They are *aLife*. The character of Life, argues Simmel (1916/1917), is *that it is and remains fragment* [sic]. Life, like nature itself, is process: a continuous flow and becoming of streams of energy in *Wechselwirkung*—interaction—through time and space. The crucial aspect of a change that complements the fragment towards a totality is available once a conscious mind becomes aware of the processing of life. Where the world outside once was but *Kräftespiel*—a play of forces—the process of the life of mind now has content and produces matters of fact *and* mental facts that are ordered; it also introduce notions of center, of distance and difference, of parallax, and diffraction for which the objective multitude of beings has no analogy. Order and center are produced to provide reference and to make sense, lest experience be only hollow form. The Kantians, such as Simmel or Foucault, know this and its consequence in positing an "I" and a "Me" as form or as carrier of the process called (subjective) experience or *subjectivity*; the "I" is an action/agency in performance, and it is, therefore, only when and where [situated] action is. And where action is, the "I," even Life (*aLife*) itself, is in a sphere where elasticity, play, tolerance, and contingency exist: Completion of fragments is the intersection of *scapes* (mindscapes,[4] somatic scapes, etc.), of topographies that delimit the negotiable boundaries of the fragments of life. In social interaction, when negotiation of boundaries fails, it is because limitations of embodiment[5] apply. Mauss enters a similar notion when he catalogues how the notion of "person" has increasingly comprised "flesh and blood, substance and form, an anatomical structure" (Mauss (1938 [1985: 2]). Cooperative interaction, as Chester Barnard (1938) urged, has the task of overcoming limitations. As such, the cyborg is the interactive totality that integrates the whole and the part through an attitude of cooperation (or mode of multiculturalism), without being either a reduction or an embroidered just-so story. As a cooperative strategy, *the cyborg is necessarily modest*. Both Foucault and Simmel can stand as *chiffres* for such a "principled defence of the autonomy and integrity of the subject" that strives to make the move from "science to culture

---

4   The concept of "mindscape" as integrated into the social brain–mind model from Chapter 2 sufficiently addresses the concerns raised in this section, but we recognize the value of iterative and multiplicative discourses that may approach this problem from other perspectives. Therefore, we revisit these questions in a different formulation both as a way to emphasize their centrality to issues of *ScienceCraft* and to engage a diversity of audiences.

5   Original text in French:

"Car, ce à quoi je vise, c'est à vous donner, brusquement, un catalogue des formes que la notion a prises dans divers points, et à montrer comment elle a fini par prendre corps, matière, forme, arêtes, et ceci jusque de nos temps, quand elle est enfin devenue claire, nette, dans nos civilisations (dans les nôtres, presque de nos jours), et encore pas dans toutes."

and from culture to praxis" (Vandenberghe 2001: 479). This means that the theory of knowledge, whether sociological or philosophical, is always relational, practical, and political; Vandenberghe identifies such an insight in the work of Cassirer. More importantly, he asks: "What is the relation between Leibniz, Kant, Marx, Durkheim, Mauss, Lévi-Strauss, Bachelard, Simmel, Cassirer, Mannheim, Elias, Bourdieu and Bhaskar? What do they have in common? What are their 'family resemblances'? They are all structuralists of sorts" (2001: 479).

Michel Foucault was also a great admirer of Cassirer. And even though he very rarely mentioned him explicitly, Cassirer is an esoteric[6] figure in the background of Foucault's writings. But in a case such as his appraisal (1966) of the French translation of Cassirer's *Philosophy of Enlightenment,* Foucault chose his words of praise as carefully and wisely as did Simmel his: Against the "nationalism of the Fascists" (Foucault), against cultural pessimism (Simmel), and, perhaps, against mechanisms of exclusion in the Western colonial matrix, Cassirer found (paraphrasing Foucault) a *calm and irresistible power, comprehending all theoretical worlds, opening the opportunity for a new history of thinking.* In identifying the *parting of the ways* that was initiated in the history of thinking, beginning in Davos, Switzerland in 1929 and culminating in the *Rektoratsrede* of 1933, Michael Friedman (2000) identified how the reductionist idea emerged as sovereign (and later moved to America) with Carnap and how the contemplative fictionalism into (social) phenomenology solidified supreme with Heidegger, leaving Cassirer to become the subject of an episode of collective amnesia with lasting consequences. However, this episode might perhaps finally be over—at least within the philosophical, sociological, and postcolonial study of science and technology. Along with an "always looming Renaissance" of Simmel and the recent "discovery" of Gilbert Simondon outside of France, Cassirer, too, is the target of a rediscovery and revival. Indeed, along with the "rediscovery" of Levi-Strauss in history and history of science (see the discussion in Breidbach 2011), it seems that this trifecta—Simmel, Cassirer, and Simondon—is poised to provoke interesting intra-theoretical developments in science and technology studies (cf. Adkins 2012, Hoel/van der Tuin forthcoming, Skidelsky 2003).

Subsequently, Marcel Mauss' affirmation of the *total human* is not a constitutive aspect or a school that we adopt and follow blindly; it serves as an inspiration within the contexts of our critique. It is, however, not that far from the idea of *natureculture*: Mauss and Simmel speak of the relation between whole and parts that cannot be reduced to the other; Mauss and Foucault speak of practices and of the techniques of the body; Mauss and Cassirer assert that the human being is also reliant on the use of symbols; finally, Mauss (and perhaps Barthes) finds *that amongst the sociologists it was the linguists who were the first to understand that all social phenomena are simultaneously and univocally physiological and psychological: Body, mind, society as psycho-physical total complexes that are*

---

6   We play here with Leo Strauss' distinction between exoteric and esoteric writing.

individuated into manifolds of mentalities that exist in the anticipation (attente) *of each other* (Mauss 1924),[7] and *Being-for-the-Other* (Parkhill 2009). Laura Hengehold (2007, also: Stingl 2011b, 2013) similarly builds on Kant, Foucault, and Deleuze when she speaks about the human body and its phantasmagoric relations. As a *heterotopia*, an under-appreciated conceptual tool from the arsenal of Foucault, the body is part of a complex set of human relations that she analyzes as part of a *negative anthropology*:

> Such an anthropology provides a necessary grounding in contingency for any Maussian notion of anticipation. Negative anthropology cannot and should not be able to 'anticipate what kind of bodies and psychology we will have in the kingdom of ends', nor can or should any 'movement or state [...] know a priori whose being will fail to benefit from a changed society'. (Hengehold 2007)

She also describes an interesting figure of the *mad critic*, a sibling to the *daft punk* or the *veiled(!) modest cyborg*: it is a figure of critique who is "not necessarily insane, but lacking the resources and the ability to communicate with a public that could have made the conflict in her imagination real and therefore capable of being altered" (Hengehold 2007:3/4). Thus they are "unable to convince anyone that their acts and words correspond to reality" and they are rendered "incapable of making meaningful history" (ibid.: 5). The *mad critic* is, obviously, an *excluded third*, a *silently Othered*, crying for a novel mode of integration such as we have argued for in the previous chapter through the concept of *narrative empathy*.

The task of this *science-crafty* kind of anthropology lies in keeping alive and unhinged the potentials to imagine and criticize, both in the actual present and in the future, the forms of encoding, embodiment, and enactment of rights and publics. The total human being as a *heterotopia* therefore encodes, embodies, and enacts the following set of relations or *phantasmagoria*:

Body <> Political Imagination <> State

From our point of view, this means that the State and the body are both formations produced in and by techno-scientific practices (Carroll 1996, 2012); in other words, they exist as *figured worlds* (Holland 1998), which can be said to be "a politically infused culture that shapes cognition as well as action. The culture consists not only of a constellation of ideas, but also physical forms systematically infused with meaning" (Mukerji 2010: 406). Perhaps State and body are indeed Simmelian totalities, and perhaps we must be concerned with their upholding a minimum degree of integrity (instead of an identity and its politics) between the

---

7   In cases where we did not have an English translation available, we took the liberty to produce an English paraphrase from the German translation comparing it with the French original.

disciplines that wish to study them. What is at stake, therefore, is the process of integration among the scientific disciplines, within science and scholarship, and between science and its publics.

In two processes that Stingl (2010b), in following Donald Levine and Paul Starr, has tentatively called "hyperspecialization" and "hyperuniversalization," we can see that scientific disciplines/schools/theories drift further apart rather than closer together. The oft-repeated public calls for interdisciplinarity are not put into actual (techno-scientific) practices. An ever-increasing number of techno-scientific, epistemic, and objectivity communities exist as separate, isolated islands in the sea of the so-called knowledge and information society.

We do not wish to attack the integrity of any one of these communities. We claim that in order to maintain their integrity, they will have to give up their isolation. We argue for a process of integration—a mode of scientific multiculturalism—that does not lead to self-destructive universalization or centralization. We argue for a mode of integration that is at the same time pluralistic, democratic, and autonomous. Scientific communities must retain their individuated integrity while remaining able to communicate and cooperate with each other. This level of integrity and communication we call *ScienceCraft*.

In order to accomplish this communication and achieve the multicultural mode of integration, science and scholarship have to solve a number of crucial problems. We do not offer an exhaustive list, but instead focus on four problems that stand out most prominently in our discussions. In our techno-aesthetic sensibility, otherwise known as a sense of humor or Wittgensteinian wittiness, we have named them after the *Daft Punk* song: the *harder* problem, the *better* problem, the *faster* problem, and the *stronger* problem. We will describe the core of each problem to acknowledge its significance, but will leave the challenge of solving them to the future. We will, finally, conclude with a possible posthuman point-of-view that will allow such resolutions to take place and occupy time, to *become,* and to account for the necessary gender critique. This will be the notion of the *emergent Third*, which we see figured in the *Three E*s: encoding, embodying, and enacting the *modest cyborg*.

The Harder Problem

It is easy to see that we have taken an inspiration from Owen Flanagan's book, *The Really Hard Problem* (2007), and David Chalmers' distinction between "hard" and "easy" problems. The *harder* problem that *ScienceCraft* will have to resolve begins with the so-called *binding problems* and extends through the issue of how social institutions obtain efficacy and agency in a material world.

The discussion of a *binding problem* originates in William James's *Principles of Psychology* (1890) in the chapters on "Mind Stuff Theory" and "Attention." The problem lies in the existence of a) an outside material world, b) the neural network, and c) our experience of and as a (non-material) *mind*, and arises in two formulations: 1) how perceptions of *a* and thoughts and ideas about *a* in *c* are bound together, and 2) how the perceptions and multifarious inputs in *b* are unified in *c*. Flanagan identifies as "the hard problem" the difficulty of explaining how mind (or consciousness) can emerge out of the materially based activity of neurons, but he raises the stakes with a harder problem: how to explain meaning in this material world? (2007: xi). We think it is even *harder* then to ask, in light of the question of meaningfulness, how social institutions, which exist in the outside world but are only partially material (if at all), can have efficacy and/or agency.

Both these problems, taken separately or together, require levels of width and depth of discussion that go beyond the scope of this concluding chapter. However, we want to argue why they are here to stay, and both problems touch upon the discussion of the idea and concept of truth (Cappelen/Hawthorne 2009, Lynch 2009, Merricks 2007). For starters, *1* is the problem of how thought and world, thought and body, body and world, and thought, body, and world cling together in a way that is possible for the reality or realities we experience to exist at all. The other problem, *2*, is formulated in alternatives of what can count as social reality as the consequence of *1*, and these alternatives fall between reductionistic materialisms/physicalisms, social ontologies such as external realism, and so forth.

Some of these issues are already found in Kant's distinction between noumenon and phenomenon, and they became the subjects of intense debate in the concept *hiatus irrationalis* in Post-Kantian, Idealistic and Romantic philosophy between scholars such as Fichte and Jacobi (Beiser 1987) and in early sociology between philosophers of the social and of science such as Lask, Weber, and Rickert (Bakker 1995, Oakes 1988). And they have remained with us ever since. A good example to illustrate the pressing necessity for an integration in the mode of multiculturalism between scientists and scholars begins with us, the authors, who disagree on how useful and valid John Searle's ideas and contributions are to the questions we address. While Stingl, despite a soft spot for Derrida, is otherwise more willing to accept Searle's external realist account of social ontology (summarized for example in Searle 1998 or McDermid 2004), Restivo is not convinced by Searle. Restivo has some well-founded reservations about Searle's theory of mind (see Chapter 2), and would certainly like to argue that when talking about social institutions and collective intentionalities, we don't need Searle when we have Durkheim. Weiss, on the third, cyborg hand, chuckles modestly when listening to the menfolk bicker and notes parallel Orientalisms in Searle's "Chinese Room" (1980) and Latour's "it's all Chinese to me" remark (Latour 1979).

These stock-debates notwithstanding, a minimal level of agreement is possible when arguing about the stakes, the most salient of which is the concept of truth. In accepting that agonistic engagements (Mouffe 2005) and advocacy di/trialogues can constructively teach us something, Restivo, Weiss, and Stingl not only opt for relational and procedural concepts and ethics of becoming, but also argue that any solution must be inherently a new version of pragmatism, perhaps even *Nü-*

Pragmatism (Stingl, forthcoming). Such an approach suggests that even scholarship that seems as far apart as that of Searle and Foucault can find common ground, as illustrated by Prado (2006: 172):

> What seems to emerge, then, is that the truth about truth must lie somewhere between the extremes marked by Searle's and Foucault's position. Most simply put, it seems that some true beliefs and sentences are true in virtue of the disposition of the world, and some true sentences are true in virtue of their status in a discourse.

For all the considerations about knowledge regimes as well as local and situated knowledges, we still find it is true that human beings like to feel a kind of certainty that they can call truth; from Kant's Copernican revolution onward, via Nietzsche, Foucault, and STS, the uncanny feeling that this certainty cannot be guaranteed has led to a lot of resistance. We tentatively claim the following about this notion of certainty:

> We have information and we have knowledge, and these are two very different but related matters. Bits of information (or data) are contained in knowledge, but knowledge is not contained in information.

Information by itself is not knowledge, and it is through the inclusion of information into actual, local practices that they are made into knowledge. Both the intelligible experience and communicable expression of knowledge by individual actors (subjectivities) tends to include the use of values.

In a very simple example, an indiscriminate pain (pain receptors and nerves firing) is a signal; it is information. This information may then be rendered into different kinds of knowledge. What this knowledge looks like, what the experience of suffering, treatment, and relief is shaped by, depends on whether it is situated in a Western medical system or in an Eastern one, such as in traditional Chinese or Indian medical systems. In both systems the pain can be made to stop, even though the Western system generally assumes an allopathic (oppositional) model for the symptoms while many Eastern medical systems utilize more holistic (in the literal sense of wholeness) or even "same-suffering" (the term would be "homeopathic," but we proceed with caution here)[8] constructions to address root-causes. Between

---

8   It should be clear that we use these terms not in the way that they are used in lay-terms but with regard to their literal and historic meaning. Nowadays people have, unfortunately come to accept the idea that homeopathy refers to a regime of serially diluted water solutions and sugar pills—a technique invented, supposedly, along with the distinction allopathic/homeopathic, by a man called Hahnemann. Obviously, we do not want to enter too deeply into discussions about anecdotal evidence, placebo-effects, or fringe sciences. This debate ranges from the truly unscientific and ill-fated "criminalization" of alternative medicine (Adams 2002) and the problems of evidence-based medicine (Cohen 2002) to

information and the different knowledge regimes, truths are produced. This does render truth uncertain, as uncertain as the differing knowledges. Truth, it seems, can be "many and one" at the same time (Lynch 2009). But the information is certain, even if it is not reliable knowledge or corresponding truth; nor is it identical to either knowledge or truth. This means that in the search for certainty, human beings when talking about the truth do indeed have it both ways; there are elements that are certain and practices that are not.

For our ends and purposes, both of these problems *1* and *2* together can, of course, be re-conceptualized in the frame of the mind/body-problem: How does a material body in a material world relate to a mind, which is experienced as non-material and which interacts with entities, agents, and/or institutions that are non-material?

Again, we consider Laura Hengehold to be our natural ally in this discussion based on her argument that both Foucault and Kant develop explicit "ideas of embodiment and imagination [… *with Kantianism arguing*] that the feelings and imaginative acts involved in pure aesthetic judgment give us access to an aspect of the body that precedes and exceeds empirical and introspective knowledge of the body" (2007: 8). As a consequence, the *body* and its imagination with which we are concerned—as a relation of aesthetic judgment, also a somaesthetic topic (Shusterman 2008)—are in-themselves and of-themselves not conceivable (introspectively) nor (empirically) operationalizable outside of the process of conceptualization. Therefore, they become a theoretical (aka, philosophical) problem for any kind of research or intervention at least in the aspect of how to connect them to any set of practices (biomedical research and clinical included) because these are not self-referential in the sense of leading to solipsism. These practices always refer to and affect the real human being who exists in her/his own life with all its references to and all affects of body and imagination. In other words, the body and its imagination is, first and foremost with regard to practices, action, and agency, a semantic problem.

Following the example of Collingwood, Guiseppina D'Oro (2007) asserts this option of a semantic solution for the mind/body problem. In light of the recurring failure to fully reduce the mental to the physical, D'Oro argues that between the

---

the serious discussion of how to conduct and evaluate trials of Western and alternative treatments (Sehon/Stanley 2003, Sehon 2010, Ionnanidis 2005).

We have little interest in quackery, nor do we wish to fall in with those who would like to criminalize any kind of alternative medicine. On the contrary, we would argue for terminological precision that accounts for actual conceptual histories, for sound research, and for exposing sham treatments and quackery of any kind, which includes both quackery in both standard Western medicine as well as in alternative systems of medicine. In this regard, we call on the philosophy and history of science and medicine to rescue useful concepts, such as allopathy/homeopathy, from quackery. When a practice qualifies as quackery, we mean to say that what is questionable is not the knowledge or the truth but the information.

two versions of the challenges to reductionism that arise—the epistemological[9] and the ontological[10]—the gap between body and mind cannot actually be conceptualized in this manner. Instead, another approach becomes necessary, an approach that should successfully account for *why* the mind/body problem is actually still here and *why*, she says, it is here to stay. This is an argument that, though it flies in the face of most established readings, manages to not just reconcile these differing accounts but to also account for their very existence. It shows how recent discussions reinvigorate older solutions of the problem through parallelist mind-brain identity. D'Oro's argument, while it might not sit well with parallelists or reductionists, asserts: Regardless of the referent(s), the discussion is not about us directly accessing different things with regard to mind and body, but "we mean very different things when we speak about the mind and when we speak about the body" (d'Oro 2007: 169). This is what makes it perpetually both a philosophical and a semantic problem that is neither epistemological nor ontological, but metaphysical in one and only one way: the semantic way. It "identifies metaphysical analysis with conceptual analysis" that is not about "pseudo-problems" or "a result of scientific advances." Instead, it is a means of "conceptual clarification" not ever intended to solve the mind-body problem or any other related problem "empirically" nor by an "*apriori* deflationary attempt at dissolution" (d'Oro 2007b: 178). In this regard, it is "an attempt to show why the problem is here to stay," or an attempt at problematization itself, one designed to show where and why there is a problem through the technique of *making it explicit* so that the problem can be addressed, communicated, and discussed.

Traditionally, the discourse of the issue and the question of "mental" versus "physical" in the philosophy of mind and neuro-philosophy begins with a discussion of the concept of pain. Pain is a central component of how philosophical discourse can connect to medicine, neurology, biology, etc. In short, "pain" is the *entry wound* through which philosophy was and is still able to enter from the humanities (the scientific scholarship made by humans for humans) into the human sciences (the sciences that are about humans [representative *and* interventive]); Norbert Elias called them "*Menschenwissenschaften.*" For our present purposes, we leave open this question: should the definition of the human sciences should cover only styles of reasoning founded in biology and the natural sciences, or must it also include the social, the cultural, and/or the political sciences?

Pain *is* a problem because it can be viewed as the very watershed through which the humanities hold any influence over the human sciences, gain importance, and draw legitimacy. One of the most illustrative explications of this issue, in the language of "mentalist vs. physicalist," can be found in Richard Rorty's seminal *Philosophy and the Mirror of Nature* (1979; see in particular Chapter 2 on the antipodeans). This general dualistic formulation of the problem is the same that Giuseppina D'Oro addresses with reference to the classic discussion by Herbert

---

9   With T. Nagel, J. Levine, and C. McGinn
10   With E. Jackson and Kripke

Feigl. The problem can be also reiterated in another question, whether pain is an event or not. But the category of the *event* is crucial. It obtains different meanings and different tractions depending on what kind of description or in what kind of context it is negotiated. The discussions that affect us emerge in equally "hard" discourses between scholarships ranging widely from Heidegger, Davidson, Deleuze, and beyond.

What people will be able to agree on is that, regardless of whether there are mental states/components in play, to feel pain at all someone must have had a previous sensory (neurologically induced) experience of pain. One must be able to understand what pain is. There are only very few people in the world who would argue that people are born into this world with an *a priori*, transcendental knowledge of pain that would enable the purely mental experience of pain without the respective sensory neurological event ever having taken place. And this does not even begin to cover the question of suffering.

For any philosophy of science and medicine, and from there to philosophical anthropology, this is an important issue. And to answer the pertinent question that the impertinent Steve Fuller never tires of posing to unsuspecting doctoral students at conferences—"Where's the politics?"—the political dimension is the result of quite an "intelligent design": Philosophical anthropology is necessarily *a priori* to political philosophy. Philosophical anthropology is inherently intertwined with the notion of the political imagination. Laura Hengehold's *The Body Problematic* (2007) is a testament to this fact. In that same respect, Hengehold also reminds us that all the *"pain*-ful" headaches that Foucault seems to cause Habermas from the point of view of political philosophy are as a consequence rooted in the very problem of pain itself. In contrasting Habermas and Rawls with Foucault, she comments that Habermas was troubled that Foucault's source of alternative norms came from the body in pain (2007: 249). Habermas' critique of Foucault with reference to his description of Foucauldian resistance was both motivated and justified "only from the signals of body language, from that nonverbalizable language of the body on which pain has been inflicted, which refuses to be sublated into discourse" (ibid.). This passage referenced by Hengehold emphasizes that Habermas' concept of language is different from, far narrower, and more restrictive than that of Foucault, and that these two concepts of language are largely incommensurable.[11] Above all, Habermas conflates information with knowledge in a way that neither we nor Foucault (we think) do. Secondly, like Luhmann, yet unlike Parsons, Habermas has little to no place in his theorizing for the body (Bare forthcoming, Stingl forthcoming). Thirdly, for Habermas, pain is intelligibly the same as suffering and vice versa. From the second and third caveats follows the conclusion that where Habermas is mostly *mental*, a broader account of the issues he raises with Foucault would actually allow for the articulation of what we are going to call, with Richard

---

11   Although there is much to say about Habermas' misconstrual and misunderstanding of Foucault, it would go beyond our scope here, so we focus on this particular aspect in this discussion.

Shusterman, a "somatic style." Style, according to Shusterman, is deeply rooted in the expression of self and society. While we often tend to see only surface representations, and the body is often reduced to such a perspective, the origins and foundations of a body's styles of expression are hardly as superficial and tentative. This perspective also connects with the ideas found in material semiotics, such as articulated Donna Haraway's notion of the "anatomy of meaning" (1997: 14).

As for the first caveat relating to the second, we want to emphasize the notion of giving *voice to resistance*. We agree with Laura Hengehold (2007) that this is a (Kantian) problem of shaping the political imagination between language[12] and body. And, as she says, this is a process that must be construed and enacted *heterotopically*. As for the first and third caveats, we think that the analysis of the relation between pain (the sensory experience), articulation of pain in the way of meaning-making (semantic), and an observing public/polis will require us to enrich pain conceptually. This new conceptualization of pain we construct in the form of a kind of "narrativization of real events" (to borrow a term from Hayden White 1981). The narrative concept we obviously prefer is the concept of *suffering* (see Chapter 5).

Pain, one might say, is rather relative insofar that it doesn't tell us much of consequence. We consider the concept of pain not as something that adds much to the analysis of events or reasons. In our view, pain is "only" and merely sensory input. It is only through suffering that pain is sufficiently enriched to attain conceptual, intelligible, and communicable importance. This is true even if without pain there would be no suffering. Pain/information can exist without suffering/knowledge, but can suffering/knowledge exist without pain/information?

Suffering, as aptly described by novelist Iain Banks, ought to be understood thusly (with substitution of "mind" for "brain" per Chapter 2):

> [T]he ability to suffer was what ultimately marked out sentient life from any other sort. They didn't mean just the ability to feel physical pain, they meant real suffering, they meant the sort of suffering that was all the worse because the creature undergoing the experience could appreciate it fully, could think back when it had not suffered so, look forward to when it might stop (or despair of it ever stopping—despair was a large component of this) and know that if things had been different it might not be suffering now. Brain required, see? Imagination, Any brainless thing with a rudimentary nervous system could feel pain. Suffering took intelligence. (Banks 2004: 279)

Often suffering and pain are used interchangeably by philosophers, sociologists, historians, even by biomedical practitioners. However, they are not interchangeable. The aspect described in the above passage from Banks makes explicit the concept of pain-related suffering in which we are interested. We argue that it is to this

---

12   Broadly construed and not reduced to language meaning only "written/spoken words," but language as sets of communicable practice.

conceptualization that philosophical anthropology and political philosophy should refer. And it is to this formulation that the philosophical anthropology and political philosophy of Kant, Jaspers, Arendt, Foucault, Bourdieu and Hengehold do in fact refer; Habermas attempts this but fails.

The following caveat could be constructed: Do children/people with ADHD or those who are deaf "feel pain" as a mental state—not physical, for sure—when confronted with the effects of their "disorders" (we mean "disorder" here in the sense of "out of the social order that one's society constructs as normal")? Surely we can imagine a situation where someone, a parent for example, may say something like "it must be painful not be able to hear the voice of a bird," "it is painful to watch my son not making friends at school," or "my daughter is in pain over the fact that no matter what, she can't get a good grade in math," and so forth. Of course, nobody would confuse this to mean the same thing as "he feels the pain of the needle in his arm" or "she is in pain from cramps," and, of course, no philosopher would blindly assign the same spatio-temporal properties to these different conceptualizations of pain. However, it shows at the same time that a folk-psychological point of view constructs the experience of what is more generally suffering on the basis of the experience of physical, somatic pain.

In this regard, we suggest we'd do well to not forget that suffering has, if not a physical aspect, definitely a *somatic* one; for example, we often let our head hang when suffering grief or stress. Meanwhile we should also strictly differentiate between pain as merely part of the somatic dimension, whereas suffering exists on different dimensions altogether (semantic and/or narrative) even if it can be grounded in somatic states or expressed somatically. As such, we immediately touch upon the issue that somatic information and somatic knowledge are social, and that somatic information and somatic knowledge about illness and disease enter into regimes of healing practices that know values such as sick, (ab)normal or healthy that transcend the efficacy of the material world but exist as material semiotics.

Therefore, pain is information and suffering is knowledge rendered from information. However, suffering is rendered from somatic information that is not exhausted by pain. With this canvas, we want to suggest that the hard problem itself, such as found in the problem of pain, will not be solved if we do not come to grips with the *harder* problems such as suffering.

The Better Problem

The following two problems, the *better* and the *faster* problems, are inherently related by the fact that they involve the organization(s) of society. They seem to be about how social practices, whereas the *harder* problem seems to determine and is affected by social practices, and the *stronger* problem seems to deal with vital materialities, material vitalities, and material semiotics more directly. We say "seems" because these four problems are intended to serve as heuristic guidelines for four fields where interested and invested stake- and stock-holders—the epistemic communities of scientists, scholars, policy-makers, concerned publics—must continually negotiate how knowledge is supposed to be formed, applied, valued, restrained, and otherwise arranged.

Knowledge regimes rest on or are composed of techno-scientific practices, so here we focus our discussion on organized practices and practices of organizations such as universities, (health) care clinics, or research institutions. In recent years (the end of the twentieth century and the first part of the twenty-first), these organizations have been increasingly evaluated by a trope from the realm of business administration: *best practices*. Talk of best practices obscures several problems while simultaneously creating more. A major issue is the disconnect between devising best practices and implementing them successfully by making sure they are actually followed. While the emergence of such mechanisms is difficult to explain, we can at least understand their implementation through the use of normalization theory, which refers to the embedding of routines and practices in everyday life through interaction chains (Collins 2006, May 2007, May/Finch 2009). This helps to reveal that the main issue we face is less the implementation itself and rather the resistance to adopting and implementing these practices. This resistance, which is not just found among lay publics but often among researchers, scholars, and techno-scientific practitioners, exists both openly and tacitly due to habits, conflicts (with *best practices* and situated propositional knowledges), or inertia, among other reasons.

Another problem arises from the use of the word *best* itself. From a semantic point of view, the word promises finality: These practices, the word itself suggests, are not just good, they are not better, they are *the* best ... end of story. The teleological closure that *best practices* include is often absolutely fatal because once you have implemented "the best," all room for further improvement has been closed off. *Best practices* signal to their users in organizations that they do not need to better themselves or their organizational procedures any further. But this is not just a play on words; from a Wittgensteinian or performative point of view, we can say that *best practices* constitute a form of epistemic imperialism. Imperial regimes of best practices, such as those currently dominating the business administrative management of science and scholarship, are counterproductive to the idea of effective(!) development of science. Our critique then aims at the following: that both in vocabulary and meaning we need *better practices* and not *best practices*. We need regimes of practices that leave room for future improvement and change, keep alive the awareness of this potential, and are adaptable to tacit, local, situated, intersectional, and standpoint knowledges.

The negative impacts on science are apparent to anybody who is currently participating in the economy of grant proposals and its regime of best practices, a regime that begins with the traditional problems of how to create research proposals. We can consider five main issues here: the hypothesis method, the order of fit problem (Anscombe 1957, Searle 1985), the positivism conflict (Adorno et al. 1976) and the *Werturteilsstreit* (Max Weber), as well as the Understand/

Explanation Controversy (Apel 1984). Along this line of inquiry, we can summarize that one can do research in three basic modes: a) beginning with a question, b) beginning with an answer looking for a question, or c) beginning with a possible answer for a question you are testing. The last we shall call the *hypothesis method*. When it becomes the adopted practice to treat c) like b) and make the problem fit the answer—and in contemporary "research management" it often does—we find that the hypothesis method is accepted as the single exclusive *best practice* without any question being raised or even allowed for being considered "irrational." This kind of *best practice*, however, is highly reductive and contorts an ideal of objectivity into a constitutive phantasm.

Phantasms

Phantasms are efficacious, dynamic, and procedural social actants, and they may be more fundamental than metaphors while bringing the same kind of constructive force as do the "metaphors we live by" (Lakoff/Johnson 1980). Two main phantasms are the "control phantasm" (the idea that human beings can gain total, mechanical control over all variables in a specific field) and the "regionalization" or "interioralization phantasm" (the idea that every entity or process can be made transparent and compartmentalized into variables)—both phantasmic processes have been archetypically made explicit in Michel Foucault's *Birth of the Clinic* (1989/1963). As an integral component of how social ecologies and figured worlds relate, phantasms, as we understand the concept, obtain structurational power over our individual and collective lives and institutions.

Our definition of "phantasm" follows closely from Ernst E. Boesch (2002), whose ideas on symbolic action establish a very precise and useful concept of phantasm. At the center of some of his work is the "myth of lurking chaos" that rules much of human civilization. A myth, he says, is a pre-structural guiding pattern and, therefore, not even a theory or precise idea; it is an "unspecified 'mould' of receptivity and evaluation" (Boesch 2002). There are different ways of dealing with myths (the myth of lurking chaos being one of the most primal and influential), and phantasms constitute one of them. Though Boesch sees phantasms emerge in the individual development of children through selection and amalgamation, his description works rather nicely to characterize patterns of phantasms generally, regardless of whether they emerge in individuals or proceed collectively. What we call phantasms are the perceiving, transforming, as well as anticipating images bound up with the acting party (or actor). Boesch, in a Lacanian moment, declares that "phantasms are, of course, 'over-determined.'" They provide a way in which culture certainly influences the way we [we add here: *including us as scientists*] think and evaluate, shapes our action interaction. However, it acts no less below the surface, in those mythical dispositions, which we now hardly notice. Culture, then, makes us form phantasmatic orientations of which we recognize the more "rational" manifestations—our goals and fears, affections and antipathies—but which nonetheless act at a depth that we will hardly ever be able to reflect on (Boesch 2002: 134).

The Two Axes that Orient Research

Notions of a global epistemic culture (Knorr-Cetina 2007) of science conflated with the epistemic imperialism of both the Western colonial matrix and best practice of the *scientific management of science* led to a scientific process that is easily controlled and disciplined in a technocratic fashion. The problem, we must stress here, is not that either of these different types is inherently problematic. Quite the contrary, each approach has its place. The problem is the lack of integration and focus on one specific type of research practice while all others are subjected to mechanisms of exclusion.

Therefore, the first premise that must be investigated is that research is oriented by two axes or binary oppositions. These axes serve as ideal-types for the study of the "direction of research fit":

a) Research can be discipline-oriented or problem-oriented.
The type we call *discipline-oriented research* follows an established set of methods and theories that are summed up under a discipline or sub-discipline. They form the *conventionalizations* (Bloor 1976) or constraints towards the development and execution of a research-project. For example, a researcher working in the sociology of ethnicity and family is only allowed to apply certain kinds of methods and to address certain types of problems. S/he isn't allowed to find any real world problem and then look for suitable methods of addressing its study and/or resolution. The only problem/s she can find and will be able to study are those that can be formed within the frame of reference of the allowed methods and concepts of sociology of the family and ethnicity. This type of research can be called *Technology* or *Engineering*, because it is an orientation of applying a standardized set of methods and ideas towards a problem that can be constructed from within those confines.

Problem orientation refers to a practice wherein a researcher identifies a real world problem, makes it explicit, and only then seeks methods appropriate for the study of this problem. In doing so, a researcher acts without the necessity of disciplinary boundaries that constrain his/her choices. Michel Foucault is an excellent example of this type of research, which we shall call *Scholarship*.

The distinction between researchers, as they self-identify, should run along those lines: whether they want to be *technologists* or *scholars*. The ideal researcher should certainly able to fill both roles. But the best practice approach that we criticize seems to constrain this option through an implicit bias in current education and research administration towards discipline-orientation and the "training" of technologists rather than the "education" of scholars. In our view, good (higher education) should establish a quality in the mix between training and education. Best practices, however, can only establish and implement practices that focus on training, since genuine education is not quantitatively measurable in the same way, if at all. Training and discipline orientation, however, equips a researcher with only limited means to acquire further knowledge outside of the discipline and the set of methods. Although an easy analogy could be drawn to factory workers trained to operate assembly-line machinery that presumes materials are already processed in a certain way, it is not merely analogy, but straight comparison when we reject the fundamental differentiation between material and social technologies: the training within disciplinary orientations is the social equivalent of factory machinery and tools. If higher education intends to equip students with applicable knowledge for the contemporary job-market and a very clear cut career track than that is fine. But it does not equip students for the future job-market or alternative careers, not to mention anything of the promotion of civics or of future challenges in science and scholarship.

b) The second distinction is between explorative and affirmative research.
Explorative or reflective research is the kind of research that does seek to discover something about which little is known or little is determined. It does not seek proof but answers questions in an open-ended fashion.

Affirmative research, on the other hand, seeks to find or construct proofs for an answer or a solution already in existence. Affirmative research is often (though not always) quantitative, since there is a culture of "trust in numbers." It is much harder to create projective research quantitatively, for quantitative methods usually need "to know what they're looking for."

Policy Research is often affirmative research, with the researcher being employed to "affirm" a policy decision that is "already been made." Most organization and evaluation research is found in this field.

Administrations and bureaucracies require affirmative research because their complexity has led to path dependencies of policies requiring only *post hoc* legitimization with their constituents.[13] In this respect, semantics (metaphors, practices, &c.) have been created along those lines. Over time, research administration, including research funding, has adapted accordingly and by the process of *virtualization* (see below on the *faster problem*); this type of semantics has spread into the process of the creation of research proposals throughout various disciplines that are far from social research. The evaluation of research proposals is, in other words, con-figured by the semantics of the affirmative, discipline-oriented rationality. In other words, most researchers are part of a global epistemic culture or figured world where imaginations are constrained to make the problem fit the bureaucracy of the research design, instead of making the research methods fit the problem.

---

13   This also fits the diagnosis of post-democracy.

Demands and Support

In Germany, it is an oft-repeated proverb of education and research that society as well as individuals must follow the idea of *Fördern & Fordern*. This call to "Support & Demand" or "Aid & Challenge" can be both used as an empowering practice or as a gesture of exclusion. Often enough, support comes with demands to follow best practices, which leads to path-dependencies that are detrimental to the actual research effort. Thus it could be that support (*Fördern*) is actually inherently constraining and, in its present state, even prevents good education and research from happening, whereas making demands (*Fordern*) provides enabling elements of freedom, spontaneity and creativity.

Actual demands *(Fordern)* or challenges are actually not put before students or grant applicants at all, because demands of actually achieving new knowledge and ideas, challenges to develop research outside of the box, are discouraged by the best practices established. For example, it is apparent from countless individual statements on this matter that the culture of "publish-or-perish" and the peer review system are dysfunctional, but there are no serious efforts underway on the system level to replace or seriously improve either. Such efforts of reform, as we will discuss in the "faster" section, must come from outside of the system lest the investigation be constrained by the very constraints it is meant to examine.

The problem with support (*Fördern*) is that it is structurally condensed to certain support-pathways and thereby creates a figured world of grant and proposal writing that has, in the meantime, led to the establishment of professional grant writers who specialize in specific grant schemes. Moreover, one cannot apply for a grant with a question in mind, but instead must seek to give a predetermined answer, one obscured in the "rhetoric of hypotheses," and provide mere affirmation without leaving the confines of that particular disciplinary sandbox.

As for demand, or making a true demand of an applicant or a student, we must enable the applicant to give an answer. A true demand puts before the applicant a question or a problem that s/he has to find an answer or a solution to without us being able to predetermine what that is going to be. Thus the applicant or student is faced with the geas to be creative and find an answer, but the reviewer must enable this by putting forward a real demand and not just a directive. If a reviewer of a grant application finds that the applicant is merely reiterating a position s/he knows, s/he should not simply say, "Well, if you seem to follow a Deleuzean interpretation of virtuality here, you have to follow in a Deleuzean project and do philosophy and subscribe to Deleuzean rather than speak about the virtuality of the body in the nineteenth century in history." A real demand is to say something like: "You may not have realized it, but your interpretation of Kant is not so new but very close to Deleuze. Tell me, do you see a way of going beyond Deleuze and surprising us with a idea or a question that shows us the matter you speak of in a novel light?" In this way, a true demand has more in common with dramatic improvisation games: the first rule is to never say "no," but to instead respond with another invitation that moves the interaction forward.

Bureaucrasia and Virtualization

The idea behind *Fördern & Fordern* in itself is certainly just as self-evident as the fact that one cannot be granted rights or entitlements without also accepting responsibilities. There can be no knowledge and no progress without both enabling and constraining. The problem with present research funding and much of higher education is that there seems to be a lot of constraining and excluding going on and too little enabling and integration. The paradox is that we need to be *less supportive and more demanding*; we need structures that enable us to make more demands. We need to ask for achievements that have quality rather than performance according to quantifiable benchmarks alone. What we have instead is the situation of a fully blown *bureaucrasia*, a structural state where individual action is so constrained by bureaucratic structure that voluntary action is the least attractive alternative for an actor who is driven into "weakness of will" (*akrasia*) by the path-dependencies of the bureaucracy—hence *bureaucrasia*.

Instead of true inventions, the products of science and scholarship are often just small extensions of already existing technologies or commentaries on commentaries—mere just-so stories. And the products of science are preferentially being reduced to cash-value and the idea of development as a growth-factor within existing technological frames. Thinking outside the box and stepping outside of the frame are acts that are treated as enemies of the tall-tale of human progress, which prizes growth.[14] This reductionism is part of the control-phantasm that has both created the hypostasis of modern insurance economies, accounting, and cost-control measures and played its part in the demise of the biological vernacular of science and consequent rise of physicalist reductionism. Alfred North Whitehead called this phenomenon "the panic of error," true progress's greatest enemy. Isabelle Stengers reminds us that Whitehead, in the final chapter of *Science and the Modern World*, has characterized this phenomenon as a danger by applying to it the railway-engineering concept of the "groove," for "the remainder of life is treated superficially, with the imperfect categories of thought derived from one profession" (Whitehead 1925: 197).

Whitehead's description fits what in other places Paul Starr or Donald Levine have come very close to; we mark this as a combination of two processes: hyperspecialization and hyperuniversalization. This combination, which from Whitehead's quote might be named "superfission," we will call *virtualization* (see below). The evasion of openness of the sciences and disciplining of all scientists (including in the humanities) to adhere to the control phantasm and stick to development and discipline-affirmative type projects that run within the path-dependencies of support structures are both the direct results of virtualization.

---

14   In fact, it is only certain types of growth that are preferred, a stipulation that will be discussed later in this chapter.

Efficacy, Efficiency, and Effectivity

One central problem we identify is that three distinct framing concepts—*a) efficacy, b) efficiency, and c) effectivity*—tend to be conflated together into an imprecise dimension of "efficiency" alone. We want to re-emphasize that being *efficient* is qualitatively different from being *effective* with regard to obtaining or possessing the capacity of *efficacy*. We argue that the contemporary approaches to (health) care management and reform operate within a single, impoverished dimension of efficiency, which is substituted in language only at times for the other concepts in practice; however, this is usually reduced to the *monoculture* of business practice efficiency.[15] With regard to connecting to the *faster problem*, we will stress below that the concept-in-use and practice of innovation is tied to the notion of efficiency, whereas invention/ inventiveness is tied to effectiveness/effectivity.

ad a) efficacy The concept of *efficacy* refers to a capacity that any entity, whether an object or agent, can be said to obtain or possess that has an evident effect on/in the world. We must carefully distinguish this concept from the others—at the very least to account for the materiality of biology—where it must be distinguished from form as well as from function. In the shortest version: Efficacy is the capacity or potentiality "to make a difference" in the world—"to have an effect." However, what this effect is, precisely and consequentially, is generally not specified.

ad b) efficiency The concept of *efficiency* is perhaps the concept most commonly used; often it is also taken to mean the same as *effectiveness* even though this is not the case. The calculations involved in measuring *efficiency* as a performance criterion require restriction to a specific context or subsystem that affords "a topology of normative reference," i.e. a function. Within a system of health care, for example, we can improve cost-*efficiency* by making sure that a specific medication is given only to those populations where its *efficacy* has been proven; i.e., it is given only to those people for whom it will make a difference (usually determined through some variety of blinded clinical trial). If we were to find that a specific pain-killer works only for a group of people possessing a certain gene, we will no longer prescribe it to people in pain who do not have that gene; thereby we have increased the *efficiency*.

ad c) effectivity Being *effective* can also mean that we make a difference in the world, but it refers to more than a potentiality or capacity while also having a higher level of specificity. For example, when we have a number of people suffering liver pain, we may have a number of medications that we know to possess *efficacy* with liver diseases, but only the one that proves to actually relieve the pain or eliminate the cause of the pain will be considered effective. Additionally, we must be sure to consider the conferral of values because they are not intrinsic. A bullet to the head will stop the pain as *effectively* as a medication for the liver, but it does not uphold the value of the patient's life that we would clearly want to protect even if

---

15   See "Stronger" for more on monocultures of thought and practice.

it is not made explicit. Therefore, effectivity is always only possible in correlation with but irreducible to *values*.

The Faster Problem

ScienceCraft, as we understand it, involves an understanding of agency, and understanding involves both concepts and practices. This understanding cannot reduce agency to either concepts and practices, nor simply synthesize them into a simple unit(y). Understanding agency means to "comprehend" concepts and practices. We cannot feign ignorance nor denounce theory as gibberish or practice as merely folksy, because complex realities around us (think of climate change) happens regardless of whether we fully understand it or wish it would go away. Wishes aren't horses and, as George Bataille said in *The Accursed Share*, "Incomprehension does not change the final outcome."

One such instance we have called the "faster problem": There is a tendency in current societies, a tendency we shall call *virtualization* (Stingl 2010b). What do we mean by that? *Virtualization* is a tendency in contemporary societies that we understand as the two hyper-accelerating processes: Hyper-universalization and hyper-specialization/-differentiation. These two processes are the figuring or formative forces that *realize* two phantasms that took hold of the imaginary budgets of modern political imagination in the course of the nineteenth century. Stingl has called these two the *phantasm of total control* and *the phantasm of continuous interioralization*. The discourse of research and clinical practice in (bio)medicine and health care was and is one of the fields where these phantasms, along with their respective (social) processes and tendencies, become the subject of science studies. But in our general acceptance of what we may call, in a Canguilhem-inspired move, "the non-innocence of scientific practices," we also emphasize that (techno-)scientific practices and governance practices are interlaced in figured worlds and establish worlds of context of meaning for each other. (Techno-)Scientific practices, in other words, do affect the political imagination in general and are found at the core of higher education, political action and decision-making, and so on.

We argue, therefore, for a pragmatic, integrated approach in science and scholarship: one which allows us not just to conceptualize, but to genuinely, and in a Weberian style, engage in public debates, policy-making, and expert-lay dyadic situations (such as doctor-patient interactions) while accounting for different environmental influences that range from the physico-chemical environment to the social ecology to individual bio-physical properties. We need a more extensive analysis of the effects that political debates, techno-scientific practices and governance (e.g. research ethics, health-care economics, and consumerism in health care), and *virtualization* have on clinical, scientific and scholarly figured worlds. This analysis must also include a critique of the bureaucracies and legislation that affect educators, researchers, and practitioners in light of the policy decisions that follow. In short, we must bear in mind the consequences of the ongoing process of *virtualization*:

a. *hyperspecialization*, a co-evolution of the techno-scientific profession, modern knowledge-based economy and politics, and the professions that study scientific development

b. *hyperuniversalization*, the trend to create categories for research, education, and scientific-public discourse that can be easily integrated into bureaucratic processes and the advancement in information technology.

Virtualization

Virtualization is the process that disciplines members of the medical, scientific and scholarly communities, administrators, legislators, and affected lay-stakeholders alike to become fit for the information- and knowledge-based economy of our times. As a consequence, *virtualization* does not simply change techno-scientific practice directly; it changes the figured world of science by changing the way we think about the human condition.

For example, in health care we are at risk of losing sight of actual patients and, as a consequence, dehumanizing health-care in the process. While technological and bureaucratic advances are meant to work for us as patients and people are supposed to be the ends of the system and not its means, the opposite seems often to be the result, and not because individual human actors in the system intend this. Quite the contrary, individual actors in charge often authentically express their good intentions. But systems we have created, such as the health care system, have come to normalize *best practices* that cannot adapt to relevant individual needs of patients and that only follow a rationality of accounting and business administration that has conflated scientific management and management of science under the imperative of *growth*. However, if we, as human beings, fail to make the impressive innovations that led to *virtualization* work for us, we will end up as mere means that exist to maintain the ongoing process of *virtualization*.

Of course, we do not claim that this is particularly original or novel. What we wish to argue is that reflections on current and past developments have led to a conceptualization of the phenomena in-play while very likely resembling the expressions of other great scholars. These reflections are useful in the way we compose them as a means to show that *ScienceCraft* requires us to be aware of structural consequences and to reflect on practices through the ability to share meaning as the source for solidarity and integration (see our discussion of narrative empathy in Chapter 5).

Hyper-specialization and Hyper-universalization

The two processes of *hyper-specialization* and *hyper-universalization* can be found explicated in the works of Donald Levine (2006) and Paul Starr (1992), respectively. However, one can easily find them expressed in other ways and with a more profound reference in the history of (political) ideas. For example, in his book, *Violence* (2008), Slavoj Žižek emphasizes how the Hegelian concept of "objective excess"—"the direct reign of abstract universality which imposes its law mechanically and with utter disregard for the concerned subject caught in its web"—is always supplemented by "subjective excess"—"the irregular, arbitrary exercise of whims" (2008: 13f.). This interdependence is exemplified by Etienne Balibar's discussion of violence in "virtual capitalism": "the 'ultra-objective' or systemic violence that is inherent in the social conditions, which involve the 'automatic' creation of excluded and dispensable individuals from the homeless to the unemployed, and the 'ultra-subjective' violence of newly emerging ethnic and/ or religious, in short, 'fundamentalisms'" (Žižek 2008: 13f.). This analysis allows us to engage in a constructive bridging of these two contrasting concepts.

It is noteworthy that inherently referenced processes of social marginalization have produced both hyper-universalization and hyper-specialization in ways that make us re-conceptualize the issues of globalism versus globalization (Beck 2006) from the point of view of discursive institutionalism. Within the fields of ideological/religious fundamentalism, we can tentatively diagnose a continuous process of internal fragmentation and specialization: Radical groups, such as terrorist organizations, that were originally formed around ideological cores diverge into ever more splinter groups which each create their own version of the ideology, at times proceeding to stages of infighting. At the same time, we do find an emerging class of the global poor or "global precarious" who each live in poverty or *precarity* as if it were a social structure or institution; indeed, institutional analysis would likely deem *precarity* to be an isomorphism.

Precarity as Contrasted with Poverty

First, we shall distinguish between poverty and *precarity*. Poverty should be defined by a lack of access to resources, wherein it is recognized by one group of (usually privileged) social actors that other social actors are poor (e.g. individuals, families, communities, or even [so-called failed/third-world] states) and do not have access to resources that guarantee even a basic level of subsistence. On the other hand, *precarity* constitutes a lack of security for loose collectives of individual social actors, regardless of their access to resources and symbolic forms of capital ("access to" here being very different from "possession of"). To belong to the *precarious class* does not necessarily entail a lack of means like education. On the contrary, this spectrum reaches from low- and no-skill workers to people with PhDs; these members of the *precarious* class do in fact have access to many resources. For example, they can make use of a nation's unemployment service or social welfare system to address present needs. But whether they can continue to develop their careers, whether they will have health care and other goods in the future, or whether they will drop from *precarity* into poverty is left uncertain. Moreover, they are losing their faith in symbolic currencies such as those that find their expression in phrases like "better education means a better future" or that "doing a good job will advance your career." While a growing number of people is suffering from this kind of disillusionment, social structures of the so-called "developed" Western societies, as well as a growing number of industrialized (or industrializing) and digitalized (or digitalizing) non-Western societies, still function on the myths and motifs of *progress* and *growth* that guided the Western nineteenth and most of the twentieth century. These myths have led to the paradoxical effect that whether one complies with or abandons these structures, one can be doomed to fall down the ladder and be deprived of resources or access to resources. As a result, many actors have fallen into a generally pessimistic outlook and a polemical or cynical stance towards their own future and the lives of others: "you're doomed if you do and you're doomed if you don't." However, they do not and cannot simply quit because the normative momentum drives them onward; path-dependencies and careers plans are built with fixed contingencies. One can choose between career step A or B, and this constitutes the "freedom of choice"; but one cannot suggest nor even conceive of a step C that is not already covered. But being offered or making the choice does not guarantee success, since there are other classes of actors (individual, corporate, shareholder collectivities)[16] who control access and resources directly. It is they who have also contributed to this (political-)economy that promotes both the isomorphic structures and institutions of poverty and *precarity* while also maintaining and promoting beliefs in symbolic capital, currencies and interests (Bourdieu: *illusio*), including the promotion of the Protestant work ethic. In other words, the idea of "upward mobility" is in peril, and the resurgence of aristocracy as a political structures is apparent.

Yet, this "aristocratic or elite class of actors" is also internally hyper-universalized and hyper-specialized. The structures, methods, and "legalities" in play are highly universal, while at the same time these elite actors are each very different. While they seem to talk the same talk and walk the same walk, when on the public stage, it is impossible to compare a high-ranking Chinese government official with a politician-capitalist like Mitt Romney, a tech-guru like Mark Zuckerberg, a media-mogul like Rupert Murdoch, a celebrity like the "Kardashian circus," or a talking head/pundit like Paul Krugman or Thomas Friedman.

The classic social scientific interpretation would read this situation in the following scenario: One could imagine all of these people conversing at a prominent international social event more easily than one could imagine that Chinese official socializing with an intern from China person who built up a career in politics through hard work and initiative. At the same time, it would be unrealistic to expect all of these people (including the intern) to have meaningful, integrative relations across their fields of expertise.

But it is crucial to understand that while this description is not wrong, it fails to cover the paradox or parallax here: namely, that the intern must at the same time act as if s/he was a member of this elite. S/he must talk the elite talk and walk the elite walk while also being just as specialized in order to be taken seriously,

---

16   ... and these constitute different types of agencies and rationales.

but s/he still would have no options to influence the path of their career actively or the freedom to make decisions outside of the pre-given frames. The elite subjectivity is equally problematic because it occurs at two poles of being part of the elite: the universalized conversation and an expert discourse, which have no connection, yet are simultaneous and irreducible. The only connection that is possible between the hyper-universalized and hyper-differentialized channels is a virtual relation. And here is the rub: it is not the universal discourse or the expert discourse that decides the future reality of either the elite member or the intern, nor is it a mere dialectical operation. It is in the zone of virtualization that one's fate is decided.

And this is where and why *precarity* exists: The fact that one can have access, talk the talk, walk the walk, and partake in symbolic capital and the same *illusio* does not decide or enable anything. It is an *emergent third* that creates new classes or new subjectivities that do not fit with older forms of poverty or success, but at the same time does not abolish them or merely synthesize them. Even inside academia, it is sometimes inconceivable why one person (usually white, usually male, *and* yet and openly declared feminist and equal opportunist) is showered in research grants and $10–25,000 in speaking honoraria while an adjunct lecturer (of any demographic) with a strong publication record and portfolio of competencies is trapped in a cycle of overly large teaching loads for insufficient pay that requires side jobs to barely cover living expenses.[17]

We are only now beginning to understand what processes we face here, and we are not moving fast enough to steer our social systems, specifically our education systems, in directions where emerging generations of students will be able to navigate and master the massive social, ecological, and medical problems that are on the horizon. What problems we do identify are in actuality harbingers of larger troubles to come. Many of today's complaints about health care, financial, and ecological crises closely resemble tourists' complaints that the waterline has retracted too far from the beach for them to go swimming when in fact it is the prelude to a tsunami: The complaint is virtual in the face of the real catastrophe that looms.

We must understand this process of *virtualization* and make it explicit. At present, *hyper-differentiation* and *hyper-universalization* both continue to structure global and local policy-making, as illustrated by current politics and rhetorics in both the American election cycle and the European debt crises: In neither case are the visible and audible actors doing anything other than engaging in virtual actions for virtual problems. Their handiwork does not take into account or interact with actual real people; at the same time, a large enough number buy into the virtual economies that they know are dysfunctional in the hope that they themselves might be the last one or the clever one to profit from it, in essence racing to be the

---

17   We are most certainly aware of "The Matthew Effect" (Merton 1968), but that only explains the "how" and not the "why," a more complex question of the type we engage throughout this work.

one to pull the ladder up behind them. But from a realistic point of view on living as a member in *precarity*, holding on to these myths has become like playing the lottery: No matter how well you play, the house always wins.

The Phantasm of Perpetual Growth

The 2012 American election campaigns have made this abundantly clear— "Again!"—one is almost tempted to say. Despite the rift that supposedly divides the Democrats and the Republicans, conservatives and progressives, or the Left and the Right, in Western politics they all do share a frame of mind when it comes to how they think human society works and develops: They believe that the path that society must take is the path of development through progress and economic growth. They believe that there are regimes of governance that regulate human behavior and that are an infrastructural force that lies either with the State or with the Market (or in the post-democratic account, the third agencies of corporations). Finally, perhaps out of an Oedipal envy of Nature's ease at effortlessly inviting exponential growth, they all believe that the role of science is first and foremost to promote human development through pseudo-organisms like networks and markets. And since human development is *de facto* equated with ideas of ever-faster (social and material technological) progress and ever-faster (economic) growth, the question is rarely asked, from whence do the components of such growth come and to where do the results go?

The only reason that living things can exercise exponential growth sustainably is because they are also consumed, whether through predation or decay, and their remains go back to nourish future growth—mediated through some radically different Other (such as plants and bacteria). Meanwhile, the virtual economy with its neverending growth is a monstrous dead-end of resources and energy, an unchecked tumor rather than a participant, as its waste nourishes no socio-ecological renewal processes. Yet science, including the humanities, takes as its mission the task of continuing to feed this clockwork beast so that it does not collapse on itself and bring down all of civilization; the only option is to continue on the path of growth and development. Even if literature, art, and other pursuits may not contribute directly to growth, as long as they can be recoded into forms of human/symbolic/cultural capital, sales, local factors—or at the very least be turned into a factor in a "civilizational/market hierarchy"—and above all be made administrable and accountable, they are granted an entitlement—a quasi-personal "right"—to exist. That is, as long as they do not disturb the peace with critical thinking of the kind that actually points out the flaws in the system.

Indeed, there is a recent trend among "critical scholars" to make sure that the "right" critical works are cited, while no action or change follows from the criticism. Citing critical ideas or writers other than the "accepted ones" or creating pragmatic suggestions or discussions from those outliers, particularly with interdisciplinary or "applied" incentives, is condemned as an "impure" activity in its own right and thus warrants intervention through mechanisms of exclusion. In contemporary best practices of academic and research administration and accounting, you either accept assimilation or face exclusion. One of the imperial flaws behind this practical or pseudo-pragmatic rationality is a particular phantasm or illusion that many participants in this economy seem to find quite comforting, but this flaw needs to be debunked over time even as the time to do so grows shorter. This illusion of eternal growth and the role that science and scholarship play in its favor are phantasms, and the notions of "purity" of science as well as of critical theory are guided by the same fear and myth of lurking chaos (Boesch 2002) that guides the market phantasm. The Western colonial matrix, in other words, is a phantasmic trope that has been very busy in first *dys-figuring* then veiling all the figured worlds in the same manner like a cultural thalidomide scandal of global proportions.

Invention vs. Innovation

We want to connect the faster problem with the better problem. This means that we must learn to demarcate the difference between the term "innovation" and "invention" as clearly as possible. Science politics and their publics often demand that science must "be innovative," but very often it is not clear what this actually means. At first sight, this would lead to the assumption that innovation is linked to *effectivity*. But, of course, one could always say that the same argument can be made for invention, and that these two are interchangeable concepts. However, we think that it makes sense to distinguish both concepts, tying innovation to *efficiency* and invention to *effectivity*. First of all, we want to say that both innovation and invention are based in the idea of (sets of) practices that constitute a specific knowledge&technology regime, machine, or (bureaucratic) apparatus that is considered new. The distinction between "invention" and "innovation" hinges on our understanding of "novelty status" and its relation to reasoning. An innovation, we argue, is only an innovation inside an existing (sub-)system, following in a specific and existing logic or rationale of that subsystem; it has to be novel but not uncommunicable. An innovation increases functionality and performance of a dimension in that rationale but it does not establish a new function inside the system. An innovation does not make a difference in the world, because it is not an apparatus that operates in its own rationale; it is already within a set of values that pre-define its existence functionally. It has already been valued and can therefore only be evaluated. An innovation is constructed with or because of an idea in mind of how to improve an existing (sub)system. In other words, it can only improve an *efficiency*.

An invention, however, introduces new functions that may or may not constitute a new (sub)system and/or new forms of rationale. An invention, therefore, *is* some*thing* different *in* the world even as its value has yet to be determined. The only thing it can be, therefore, is *effective* or not *effective*; it cannot be *efficient*, because *efficiency* is an internal measure to its own functioning. An invention does something different or new, but whether it does something well or could improve on the status quo is another question entirely. Therefore, one can create innovations for an invention, but not the other way round.

To learn how to compare effectivity and efficiency, invention and innovation, we find the following illustration a useful one:

> A number of people living in location A need to get to location B regularly; for example, a number of residents in one corner of the city of Los Angeles need to get to the other corner everyday to get to work, and also take the return trip home. A lot of people use their car to *effectively* do so.

A car has the *efficacy* of moving a person from A to B. In this scenario, it does make sense to build cars to be more *efficient,* such as by making their gas-consumption more *efficient,* or by making them faster (speed-*efficiency*), and as a consequence making them more cost-*efficient.* At first glance, this does also seem very *effective.* But remember: an innovation that allows a car owner to be more efficient only means that it is efficient for the *subsystem of the car.*[18] Effectivity thus comes into question when considering factors outside of that subsystem. First of all, since we consider effectivity as moving from A to B, our values may include "safety," "reliability," and "time." As long as the same number of people needs to go from A to B and the only means to get there is a car, there is little people can do; they are stuck with this particular subsystem and its constraints. But what if the number of people who need to make the move increases, or alternative technologies of movement are introduced? These commuters are no longer trapped with only affordance; they now have options, and with the options come new factors, both positive and negative, to consider. Our point is that the focus on efficiency alone is genuinely unrealistic.

We continue to build ever bigger, faster, stronger, and/or fuel-*efficient* cars. But any resident of Los Angeles (or any other major city in the U.S. or the world) who has to go from A to B will tell you about the daily rush hours and traffic jams that arise as confounding effects resulting precisely from the wholehearted adoption of the "most efficient" method. Even though fuel-*efficiency* (and accompanying waste energy harvesting devices like flywheels) are helpful in mitigating the annoyances of traffic jams, one ought to step back for a moment to ponder whether a car is truly the most *effective* way of getting from A to B. If you do not have trains in your civilization, "inventing" trains[19] could be a break-through since trains, like cars, have efficacy to move you from A to B. If you do have trains (or other forms of public transit), one might suggest that drivers

---

18   If we were Luhmannian scholars, it would more appropriate for us to conceptualize this as the subsystem of car commuting as a communication system. The authors are divided on the viability of this approach: Stingl, whose father was a car mechanic, thinks this is ludicrous; Weiss, whose father was an engineer, prefers it.

19   It could be argued that implementing modern high-speed rail in America would be a truly novel invention because it would actually make rail travel feasible.

use them instead because they would be a more effective way to get where one needs to go.

Another example relating innovation and *efficiency* would be creating a cell-phone that could store 100, rather than 12, ring-tones; it is certainly an innovation and an increase in *efficiency*, but it is not an invention, and—with regard to the function of the cell-phone—not particularly *effective*. In fact, simply increasing the number of available ring-tones could make the ring-tone feature harder to use if a better sorting system were not also included because a user would have to search longer for a tone that they wanted to use. This does not mean that it is a *bad* thing to have created a cell-phone with more memory; it simply means that it is neither good nor bad, just more efficient within that particular system and its logic while making little difference in the supersystem around it. These examples illustrate the crucial difference between these concepts, a difference that is equally valid when talking about biomedicine, health care, and care systems—sites for many investigations from science studies and philosophy that examine conceptual and ethical practices. Most reforms and novelties in biomedical practice and healthcare management are, unfortunately, very *efficient* and innovative, but not very effective. This is why we speak of *effective* treatments as cures even as we can make existing treatments ever more efficient without making them effective; we are assuming here that the *effective* goal of a medical regime is to work towards the (multiple dimensions of) health of the patient.

Finally, we must not forget that we have need for a new economics and understanding of actual real-world connections and consequences between techno-scientific practice and governance, education, society, and economics in the form of the political imagination. We can agree with many of the diagnoses that we find in Tim Jackson's *Prosperity without Growth* (2009), Richard Heinberg's *End of Growth* (2011), or Timothy Norton's *Ecology Without Nature* (2007). The case in point is that in techno-scientific practice, policy-making (in particular in regard to techno-scientific governance), and education (from mathematics in schools to academia, or in civic discourses with the public) we must emphasize the separation of (techno-)science from an arbitrary factor such as *growth*. We must promulgate the understanding that growth, by design exponential, is not sustainable and leads either to entropy or collapse. In simple words, *growth* means that development is reduced to a development that is becoming ever faster (in ever more specific yet universalizing, quantitatively measurable variables like *virtualization*), instead of deeper, wider, and more sustainable. This is, in short, the *faster problem* we have to solve, and to return to the inspiration behind this section, Georges Bataille, we want to close with his thought on this matter, which we think describes the idea behind the gesture or movement of *ScienceCraft*: "Woe to those who, to the very end, insist on regulating the movement that exceeds them with the narrow mind of the mechanic who changes a tire."

The Stronger Problem

In the previous sections, we have discussed several topics in the framework of "harder problems," "better practices," and "faster growth": better practices as an alternative to best practices, *efficacy* over *efficiency*, the increasing *virtualization* that occurs through *hyper-specialization/-differentiation* and *hyper-universalization* that leads to increased *precarity*, and the myths of boundless, unchecked growth. To address "the stronger problem," we will take as our stepping point the caution against "monocultures of thought" and apply the lessons of the previous sections to explore the disconnect between common perceptions of strength and effective strength. We reject the modernist, patriarchal cosmology that begins with "Cogito, ergo sum," and we cautiously embrace the postmodern one that looks first at the decentered connections to move on to the networked perspective that seeks to enable the actors, identified by their ability to suffer, entangled within.

Epistemic Monocultural Systems

Many types of problems we see today are a result of an epistemic monoculture in its mature stages where resources once considered free dwindle, waste products stack up faster than they can be dispersed, *precarity* has become rampant, and structural collapse is imminent. We are left limping along, yearning for the "good old days" when everyone had home-cooked meals in houses with white picket fences and people were too polite to discuss politics or religion, when men were men and women were women and children were obedient, and when the American Dream of prosperity for those who worked hard was still plausible, and Europeans hardly felt the need for a Dream of their own. It was so easy to have harmonious relations with your neighbors when your families had known each other for exactly the three generations that suffice for oral tradition-building, you all attended the same Church, the men went to the same bars and the women hung laundry on the same days. In these nostalgic communities, there was little need for hard, explicit rules because the important things were already tacitly agreed upon by everyone by virtue of common culture.

But this is merely an illusion. The costs of these systems were externalized, swept away so that no one needed to worry about their cultural waste products. For example, female sailors in the Navy who become pregnant are often separated from the service with negative marks on their records—for "damaging Navy property." Even if the father were also a sailor, the woman would be the one to bear the brunt of the consequences. Opponents to gender integration in the services claim that this only became a problem once women were allowed to serve. But there were plenty of pregnant women before; they were just civilians living in the various ports visited by ships full of randy sailors. Those women were invisible to the system of discipline; these women are made all too visible (sometimes in public Captain's Masts)—yet both are deemed "immodest."

The slow food and food justice movements run into the same gendered speedbump, albeit from the other side. In one breath, advocates expound upon the benefits of home-cooked meals for their nutrition, cost-savings, and social value, but in the next breath have to qualify that they don't necessarily advocate that women be put back into the kitchen. Many advocates don't even bother with the second half, leaving that question to float in a breeze of ambiguity. But what other option is there? The entire tradition of home-cooked meals for the family was built upon multiple normalized social practices: the private home, the 9–5 work/school day, a hearty dinner, and a wife/mother who is there to bring it all together. Even the wife traditionally had servants to assist with or completely handle the cooking of lavish meals for the family; it was only in recent times that all housework was relegated to the homemaker (with the assistance of "labor-saving" technologies that only produced more demand for labor). If one considered the normal opening hours of stores and delivery practices of courier services,[20] one can still see the embedded norms of the wife managing the home while the husband worked (and the children attended school): who else was supposed to go shopping and receive packages at home during working hours? If we will not reject the structural conservatism that keeps us dredged in these old habits, how can we expect nuclear families to magically find the time and wo/manpower to suddenly start cooking traditional meals from scratch? Even guilt, a prime motivator for many a mother, only goes so far.

There were other invisible functions of the homemaker as well. Many political campaigns were once staffed and supported by wives, who provided volunteer labor to sew sashes and bunting, feed campaign offices, and coordinate outreach; once those women found outside employment, this niche was left empty. In recent years, these less visible but highly central roles came to be filled by a combination of young activists and older people (many of whom were also activists in their younger years), but with increasing pressures on the younger generation and decreasing quality of life and employment, the younger activists have found less and less time to devote to civic participation. Conservative political organizations were able to counter this trend by channeling resources in the form of college and law school scholarships, housing assistance, promising internships, and career networking: these all went towards the new generation of conservatives in the past few decades, and this investment has paid off. Meanwhile, their liberal counterparts, often lacking the financial and institutional resources—but more importantly, lacking the coordination and vision—witnessed a distinct slump in youth participation as young activists, rather than feeling welcomed and encouraged by their seniors, were often drafted into service as low-level volunteers doing sign-waving, envelope-licking, and doorbell-knocking. These activities demonstrated a gross misunderstanding and underutilization of the potential skills and abilities

---

20   Although many retail stores have expanded their hours in recent years to accommodate working people, courier services like UPS and FedEx still offer few options for those who cannot guarantee home presence during the day to receive packages.

offered by the younger generation, who were tech-savvy, adept at multitasking, and often more specialized in subject matter knowledge.[21]

These are but a few examples of the ripples resulting from the integration or exodus of a particular type of actor from a place in society that had been normatively calcified around their absence or presence. There are many other places where similar effects can be seen in other realms and subject areas, and they all point to one main challenge: The system we have based our society on in the West does not easily yield to alternative practices.

What adaptations we have seen, such as the shift from "physician as paternal figure" to "physician as mentor" to "physician as partner," are also accompanied by other power shifts that often compromise the promised increase in agency—direct-to-consumer marketing and the patient reframed as patient-consumer, for example. This has deep implications for practices of care and the valuation of expert knowledge, as patients are urged by television advertisements to "ask your doctor if [insert named pharmaceutical] is right for you." Although paternalism as exercised by physicians ought to be continually questioned and evaluated because it can lead to gross denials of personal health agency—such as with childfree women seeking voluntary reproductive sterilizations but being denied because of age, marital status, and parity (previous incidents of childbearing)—removing the underlying value that experts have a duty to protect non-experts when collaborating has dire potentialities. And turning care and medical treatment into merely a commodity that one can "shop around" for is not the way to do it.

What we face, then, is both a failure to adapt to complex multiplicity and a failure to be able to talk about multiplicities in fruitful ways. This section will tease out three key foci that will lay groundwork for future engagements with these complex situations: 1) stronger agencies, 2) stronger subjectivities, and 3) stronger temporal-spatial engagements. These will be introduced in three converging fables: the Lion of the Tarot, the Parasitic Mouse, and the Persevering Bear.[22]

---

21   Weiss once participated in a focus group by the League of Women Voters that sought to understand why young women in particular seemed less politically involved. The findings were presented to a large membership meeting with the focus group participants up on a stage. The severe disconnect between the older generation on the floor and the younger generation on the stage was palpable as these very differences in expectations and abilities were made explicit.

22   It is true that there is a long history tracing back to the Greeks of analogizing social processes to biological ones, especially of comparing civic rule to bodily processes of control. And it is also true that there are certainly dangers in equating the two uncritically. However, there is also benefit to "following the energy" in social systems while keeping in mind both the features of ecologies that allow them to perpetuate and the disruptions of those features that cause them to stagnate and eventually fail. In the style of Serres, these animals are not animals qua animals, but instead are inhabitants of cultural fables that breathe social ideas and images.

The Lion: Agentic Strength

Let us begin with a common symbol of strength—the lion—and consider it within the symbolic context of the Tarot. The Strength card in the Tarot is usually number eight, and it portrays a maiden with a lion. The lion, bedecked in a train of flowers, is tame or even bemused as it looks at the woman, and the maiden's hands usually rest on the lion's jaws, as if to keep them closed.

If we muse on possible interpretations of this image, we can come up with multiple interpretations. Does this mean that the woman is so strong she can subdue a lion, the usual symbol of strength? Or is the chain of flowers so strong that the lion cannot break it (and thus has given up?). These are both relative measures of strength—relying on the intra-action of a known quantity (a lion) and an unknown (maiden or flower-chain)—but they do not ascribe agency to the lion. This is a standard modernist interpretation, a morphed maiden–unicorn story—with purity substituted by strength. The lion is merely the foil through which the active maiden comes to know herself: the woman is either directly stronger than the lion or she is able to employ a device (the flower chain) stronger than the lion.

But let us turn this around and consider that the statement of strength is not one of conquering or of exceeding, and that we can engage with the agency of the lion without losing respect for the maiden's agency. Perhaps the lion is just as strong as he has always been. Maybe he is so sure in his strength that he is willing to submit in this instance for the sake of connection with this woman? Maybe he wishes to help the maiden to feel strong, and so generously lends her his power so that she can feel mighty today. Joss Whedon, through *Buffy the Vampire Slayer* (1997–2003) suggests that we can share our strength with others and become even stronger together; this represents a feminist theory of power and strength that defies the economics of scarcity that normally frame these exchanges. It is through the iterative exchanges between trusting participants that potentials are multiplied, generating solidarity and connection.

In this we see a discussion between patriarchal domination, individual-effacing collectivism, and reflexive situated knowledge (our modest cyborg). Perhaps the lion, knowing that he could never connect with such a maiden unless he were non-threatening, intentionally participated in an activity of *enablement* to encourage her performance of agency. Indeed, like Jung's *anima* and *animus*, this lion draws out from the woman a deeper strength through his belief in his own security and in her potential to become strong.

The Mouse: A Strong Guest

Weiss Meets Mouse[23]
One day, I saw a mouse run (incredulously) quickly across the kitchen floor. A literal embodiment of Serres' *parasite*, this being was probably the source of the

---

23   This section is intentionally written as a narrative by Weiss.

noise I had been hearing in the ductwork of late, a scrabbling that could easily be dismissed as a noisy expansion of metal or the running of the furnace. Once this previously unnoticed entity had made itself known to me, I found myself entreated by various websites to "mouse-proof" my home, which in a very real way required me to see the world as a mouse does. Small holes, even the size of a nickel, are now important access points, a bag of flour—an ingredient with potential to become delicious—is now an easy feast, and nooks and crannies that would often be ignored in cleaning now become potential latrines. Steel wool is recommended to block holes, since it is annoying for mice to try to chew through; this cleaning tool is repurposed as a miniature barbed wire fence.

I would probably never physically come into contact with this mouse, and while I would be considered mightier than it, the fact that I have to call a pest control expert who knows its ways (hopefully in not quite as intimate ways as Christopher Walken's "Caeser, the Exterminator" character in the movie *Mouse Hunt* [1997]) as a go-between, an anti-diplomat to make it harder for mouse and me to connect, rather than easier, suggests that it is in fact stronger than me in many ways. Interestingly, many experts today label themselves as "pest management experts" rather than as "exterminators": the one I consulted impressed me with his focus on empathy for the mouse's situation. In the end, we decided to not take action since there were no signs of a takeover by a family of mice and it probably was a *nomadic* mouse. But this exchange left me with a changed perspective; I shall always see a house differently because I now know of the view from below.

The Strongest Husband

In several Asian mythical traditions, there is a story about a mouse family that had a daughter. Mother and Father Mouse, being indulgent parents, wanted to find the strongest husband in the world for their child. So they went to the Sun, whose light warmed the coldest lands and who was clearly strong enough to promote all life on Earth. The Sun declined their offer because he was not the strongest; Cloud could cover his face and prevent his light from coming through. The parents went to Cloud, who told them that Wind could move him around against his will. When the mice went to Wind, they were told that only the great stone Buddha could stand against Wind's might. But when they went to the stone Buddha, the statue sighed and said that while he may stand against wind, rain, fire, and even earthquake, someday the little mouse living under him will dig out enough dirt to make him collapse. Satisfied that they had at last found the strongest being in the world, the mouse parents took that mouse home to meet their daughter.[24]

---

24   This parable likely is based in Confucian norms that admonish people to "know their place" and to pair up with someone of comparable station in life. From that perspective, the mouse parents would be guilty of excessive pride and were humbled by the clever Sun and his friends, who taught them a lesson about appreciating what was within reach. But in the spirit of creative appropriation, another reading offers much to this discussion.

Relational Strength

These two stories show us the *relational*—as opposed to *relative*—quality of strength. Relative strength is a comparison based in domination and dismissal: "You are strong, relative to a mouse." But relationality is not about reifying different strengths and then putting them in head-to-head competition; it is about identifying the ways in which these two (or more) entities interact, keeping in mind all of the previously discussed *umwelt*, cognitive, and social conditions. Most importantly, in order to develop that relationality, both the agent and its agency must be recognized. This relationality is transformative as well—Weiss saw the world from the perspective of a mouse, and Mrs. and Mr. Mouse stopped looking beyond their own strength for satisfaction.

Parasitizing Serres' Parasite

Michel Serres' *Parasite* presents the idea of an intruding, interrupting third party on a private engagement who, through this invasion, disrupts the transaction at hand while undermining the benefit of the original two parties. This "unwelcome guest" has no right to be there, as determined by expectation or by rule established by the original two (always from a limited and limiting objectivity that excludes some forms of being). This third party is Other, foreign, unhelpful (again, based on the limited/ing objectivity of the two parties). But recent studies suggest that the presence of certain beings we would call "parasites," like tapeworms, may actually offer some health benefit in the form of reduced allergy symptoms (a *belle noiseuse* that distracts[25] the immune system from overreacting to harmless pollen proteins). Thus what we dismiss as parasites may in fact have more in common with Myra Hird's symbiotes (2009), which she emphasizes are not the completely altruistic and good-natured partners we anthropomorphize them to be. These symbiotes, she argues, exist in a dynamic tension with their partners, and from that tension is derived benefit for both (or all).[26] Less controversially, new research involving fecal matter transplants aimed at re-establishing a diverse and healthy population of gut flora (a nicer way to describe the colonies of bacteria that live inside our intestines) have proven to be so successful that clinical trials were prematurely terminated because it would have been unethical to control groups to continue (and thus prolong their IBS [irritable bowel syndrome] and other gastrointestinal woes resulting from impoverished gut biomes). No one invited the ancestors of these bacteria into our intestines—they snuck in while we transitioned from fetuses to babies through our mother's birth canal (or later, if born through Caesarean section) and made a home in our bowels—but we realize

---

25   We use this anthropomorphizing term intentionally as a demonstration of our inclusive agencies.

26   One could argue that many interpersonal relationships among humans could be uncomfortably similar to this type of exchange: romantic partnerships. professional collaborations, and even parent-child interactions.

now that we can't live without them (as anyone who has suffered GI distress after a long course of antibiotics can attest).

Here we will parasitize from Serres and derive a *parasite* with a better public relations agent: the Emergent Third. Rather than an uninvited guest, this entity is sought after: hors d'oeuvres have been budgeted, a place has been set, and an invitation has been tentatively sent. The host has even invited guests who may be compatible with this potential guest of honor. What is tricky is that it may not be clear when this special guest arrives: it could be felt as an *affect*, an atmosphere of closeness, or it could be a new idea arising from eager conversation. The Holy Ghost at Pentacost, the drink for departed friends, "rolling" at a party or a rave—these are examples of this Emergent Third, a formerly unwelcome party crasher, who fills a need we didn't know we had.

The Bear: Strength of Perseverance

In Korea's creation story, the King of Heaven came down to Earth to look for a wife among the animals (as there were no humans yet). After talking with all the animals, he settled on two possibilities: the fiery Tiger and the patient Bear. He could not choose between them, so he gave them a trial to complete: They had to spend 90 days in a dark cave eating nothing but bitter roots and herbs. Whichever one stayed the full length of time would get to be his wife. Both Tiger and Bear agreed, and they went together into the deepest, darkest cave and sat down in the blackness.

Time passed, and Bear and Tiger grew more miserable: the cave was dank and cold, and being unable to see any sunlight made it hard to keep track of time. The herbs burned their throats, and the roots puckered their tongues, and there never was quite enough to make a full meal, so their stomachs rumbled. More time passed, and Tiger because to get frustrated: "Why does he keep us down here in misery? Is he trying to punish us?" Bear tried to soothe her friend, "He is the King of Heaven, and he is very kind; I know that he means well by this. And if we both can make it through, maybe we can both go live in Heaven together! So stay with me, Tiger; we just have a little more time to go." Tiger roared in anger, "I can't take it anymore! I eat meat, and I haven't had meat for months! If this is what it takes to be his wife, I'd rather live alone on Earth." And with that, she stormed out of the cave into the blinding sunlight.

Bear wept, but stayed put in the darkness, eating her bitter sustenance. When the day came for her to emerge, she heard the King of Heaven call down to her to come out, so she did. He congratulated her for her strength and perseverance, turned her into a beautiful woman,[27] and married her.

---

27   This fable, like the one of the mouse husband, likely was intended to convey cultural norms that prized perseverance in Korean women; given the long cultural history of aggression from outside invaders, it is not surprising. And tigers appear as antagonists

The Temporality of Strength
This story alerts us to the dangers of considering strength only as an immediate effect and urges us to keep in mind the importance of effectivity over time. This leads us back to problems of measuring factors like efficiency, which tend to use simple measures of action and often omit relevant factors altogether. Additionally, sometimes inaction is the best course of action. For example, recent studies of stock traders have suggested that testosterone not only promotes greater risk taking by the traders, but that its levels fluctuate over time and in response to performance—successful transactions give a payout of a testosterone increase, prompting further risk-taking. Therefore, in order to appreciate the potential impact of this social-biological assemblage, we have to think iteratively, over time. However, we must also appreciate the strength of temporality: because we move through time in one direction, it becomes surprisingly easy to rewrite our narrative based on present revelations, such as when people who achieve a new understanding of their sexual orientation may rewrite their past based on this perspective rather than to embrace the more uncertain possibility of changing over time.

Social Tolerance, Immune Investment
Intersections of space and time also come into play when drawing connections between social interaction and health. Researchers have related factors such as feelings of loneliness to immune response and overall health,[28] proposing connections between social and physical health. However, read orthogonally (or perversely), one could also surmise that the body is less likely to invest resources into building up the immune system when the individual is not in the communal cesspool of disease and competition that is the populated social sphere. This perceived strengthening of the body's defenses against disease only happens because of spatial and temporal iterations that repeatedly put it into contact with potential threats; in order to persevere through this hostile environment, measures must be taken to balance out the increased danger and discomfort.

Fragility

Now that we have discussed three types of strength, let us consider an opposite: fragility. While "weak" is usually the opposite of "strong," and so one would

---

in many other Korean folk tales, perhaps mirroring the fate of Lillith. Once again, we show appreciation of these stories by adding another layer of interpretation to enrich its relevance.

28   See Cacioppo, John T., and Louise C. Hawkley. "Social Isolation and Health, with an Emphasis on Underlying Mechanisms." *Perspectives in Biology and Medicine* 46, no. 3x (2003): S39–S52. doi:10.1353/pbm.2003.0063; Pressman, Sarah D., Sheldon Cohen, Gregory E. Miller, Anita Barkin, Bruce S. Rabin, and John J. Treanor. "Loneliness, Social Network Size, and Immune Response to Influenza Vaccination in College Freshmen." *Health Psychology* 24, no. 3 (2005): 297–306. doi:10.1037/0278-6133.24.3.297.

presume that "weakness" would be an appropriate opposite to "strength," it is better to look at the word "fragility" to better carve out the epistemic space we want for our discussion of the "stronger problem." While weakness implies a lack of ability to effect change, fragility suggests something more pervasive—a vulnerability to breakage, an inability to endure changed conditions. We mark some packing boxes as "fragile" in the hopes that they will be treated more gently: reduced force of the hands on the package, lower acceleration rates as the package is moved in space so that the contents don't brake/break against the inside of the container. These fragile items—glass, crystal, balsa wood—are often "pure" objects, made from one type of substance and made vulnerable by virtue of their structural design (a wine glass has a bulbous top with a narrow stem, for example). So too are social norms and structures that presuppose a monocultural or monoconceptual working material.

Fragile Heteronormativity
In discussions about gender and sex identity, the term "fragile" often arises when proponents of heteronormative cis-gendering assumptions claim that "gender identity is fragile"[29] and therefore ought not be meddled with by allowing deviant behavior. But here the language presupposes many things: first, that a proper gender identity can easily be broken because it is so delicate or difficult to cultivate properly, and second, that there is a single "correct" progression that can be interfered with or worse, corrupted by abnormality. Both of these model gender identity as some sort of a crystalline structure—regular, orderly, and naturally arising from some fundamental natural characteristic—that could fail should one ion be out of place or worse, substituted with the wrong one (a doll in the place of a fire truck, for example). The same restrictions are imposed on sexual orientation in some parts of America: a man who has one sexual encounter with another man, even if he was not and is not interested in men, might be socially pressured to forever-after identify as bisexual because he has been "marked" (as Haraway describes) by that one deviant sexual act. In some cultures in America, sexual orientation is not an orientation of desire; it is a catalogue of one's past and future actions more akin to Santa's "naughty or nice" list. All of this paranoia and dictatorial control is aimed towards one goal: reproduction of individuals within the human species.[30] Although it is supposed to be "natural" to be straight and cis-gendered, it sure seems difficult to successfully pull it off in this worldview. The glimmer of hope in this anti-compassionate wasteland is the revelation that this static, sin-ridden house of cards can be rejected in favor of a performative and iterative model that embraces multiplicity and diversity; when this happens, gender identity becomes stronger.

---

29   My thanks to one of my students who asked about this common phrasing.
30   This idea is itself based on the primacy of genetic propagation over other types of reproduction, a normative assumption ironically not confined just to wet scientists but also underlying life choices of scholars who reject genetic determinism.

A word of caution: just because things may exist in a multiplicity does not necessarily mean that the result is stronger. In fact, gestalts of traits and norms, such as of "biological sex," are extremely fragile extrapolations that fall apart upon the slightest interrogation or division since no one factor is both necessary and sufficient to determine categorization. A person can have chromosomal sex (XX or XY), [31] hormonal sex (relative levels of estrogen/testosterone), and developmental/genital (testes or ovaries, penis or vagina), for example. The only common binding factor is the ideological assumption that *there must be a binary categorization scheme for people based on sexual activity and reproduction*, and when that is called into question, the integrity of the gestalt falls apart.

Normative Vertigo

However, one could take the opposite position, from a demasculinized logic such as that advocated by Andrea Nye, and argue that it is exactly that nebulousness and irrationality that makes the category so strong that it defies all "logical" evidence to the contrary. When this happens, we see an instance of *normative vertigo*, whereby the sudden loss of classificatory grounding drives the observer to actively and desperately grasp another standard; when someone identified as a woman says that she has no uterus, it is pointed out that she still urinates sitting down (and probably still uses the ladies' restroom, quite a tautology!) and therefore is still "female enough" to warrant gendered treatment.

Look at the case of Castor Semenya, the South African Olympic champion runner. Once she did "too well" for a woman, her sexual identity was called into question—revealing unambiguously that one of the motivations behind having categories based on sex in sports is to protect delicate women from men who would overrun them. And thus proceeded a physically, socially, and emotionally grueling series of degrading tests that desperately tried to apply the nebulous gestalt of sex to this one transgressive individual.

Misleading Multiplicities

Two examples of current attempts at multiplicities that have not yet reached a critical mass of critique are the labeling of efforts as "sustainable" or "interdisciplinary." Although both have been around for decades, there have been renewed efforts to raise the visibility of activities that could fit under these labels. As with the gestalts of sex, it can be hard to find the bright line at which "sustainable" ends and "unsustainable" begins, and where "interdisciplinary" actually is "cross-disciplinary," "trans-disciplinary," or merely "multi-disciplinary." Another characteristic that makes these two examples differ from our multiplicities of strength is the false assurance of concrete objectivity.

Depending on which apparatus of logic I apply, the answer, and even the question, changes. Such is the way of strength, as it always relies on something

---

31   Of course, this gross simplification forgets about other configurations of sex chromosomes, such as XO, XXY, XYY, and XXX.

against which it is to be measured. With a small change of Navy physical readiness test regulations, a Sailor could drop from "Excellent" condition to "Average" if the prescribed angle of the arms for pushups is altered, or a rule is changed to prohibit a partner from sitting on one's feet to provide leverage assistance for situps. Indeed, the very construction of the test emulates a limited male version of strength, not a human one.

Interlocking Interdependence

This is not mere numerology—we do not advocate for looking for a third everywhere. This only emphasizes the need to look for the forgotten/excluded/silenced position. This could be the "fifth quarter"—another name for "what falls out" when a food animal is quartered—offal food for the peasants that is now prized by foodies as "nose to tail" eating. Likewise, Stanford University's "6th Man Club" is a school organization that passionately supports the basketball team as fans on the sidelines. The conceit of the "6th Man" is often used by basketball fans to suggest that they are not merely spectators, but active participants in the game activity; theories of affect and co-presence would support this framing and applaud such a recognition.

It is these interlocking and interdependent agencies that offer a greater strength. A recently discovered species of shrew, *Scutisorex thori*, is thought to be related to *Scutisorex someren*, the "hero shrew," so named because it is able to tolerate an incredible amount of weight on its spine (a researcher stood on its back for a minute and it walked away alive and unhurt). This heroic feat of strength is made possible by a unique configuration of its spine: the vertebrae interlock with each other, resulting in a visual representation of networked multiplicities that support each other in an existential entanglement.

This is about adding a perspective, about recognizing the agency of a party that previously could not speak or could not be heard. And so, carrying forward the benefits of multiplicity (agential diversity), intra-activity, and iterative diffraction into the realm of knowledge practices and regimes, Weiss joins the advocacy of Haraway, Harding, Longino, and others to frame work done by feminist STS and other science studies scholars as *strong epistemology* and *strong objectivity* because these descriptors better characterize what these theories (from situated knowledges to standpoint theory to agential realism) actually *do*, especially in the context of challenging and disrupting inlaid masculinized binary segregations of knowledge. It is not intended for these theories to completely shed their feminist geneology, just to adopt another garment for other types of situations where there is less interest in inclusion, but still plenty of interest in *better* practices.

Reflections

To riff on Serres' *Malfeasance*, where he distinguishes between "hard" pollution and "soft" pollution, I distinguish between "hard" and "soft" material agency. "Hard" agency is something that is semiotically recalcitrant; it does not play well with other agencies. "Soft" agency is something that, because it is perceptible and understandable to other (relevant) agencies, does. However, the ethical evaluation of whether a "hard" or "soft" agency is "good" or "bad" relies on the context, even the *telos*.

The use of gutta percha is an example of this "hard" versus "soft" material agency; even though it is soft in texture and pliable, it is also biologically inert. These two features together—two different types of strength—make it the ideal substance for certain medical procedures, such as root canal tooth fillings. It is even more inert than metal since it does not "communicate" through ion exchange. Although organic recalcitrance leads to problems in the wrong context—consider the gargantuan islands of plastic in our oceans like the Great Pacific garbage patch—for non-disposable, non-consumable items like medical implants or similar structural supports in our bodies, we are better served when they are "harder" in their agencies. Gutta percha, then, is strong through a combination of the Lion and the Bear—it is both agentic and persevering—through its incommunicability with the life ecology of the mouth.[32]

The context-dependence of "hard" and "soft" agencies can also be seen in apparently unrelated events that may in fact be connected: the disastrous bee die-off that continues today and that was named "Colony Collapse Disorder" in 2006 (van Engelsdorp et al. 2006) and a sudden worldwide shortage in mealworms in 2008. Mealworms are an important source of food for many types of animals, include exotic pets like reptiles and frogs. The main mealworm husbandry companies had all switched wheat suppliers; subsequently, their stock failed to breed successfully. Another crash happened a few years later. Meanwhile, theories about CCD include concerns about genetically modified crops, specific types of pesticides or herbicides used on crops that attract bees, and monocultures that result in a single-source diet for bee colonies. Additionally, colonies are often fed high fructose corn syrup as a supplement to their impoverished diet, which reduces their intake of health-boosting nutrients that are present in a normal pollen-rich diet (Mao et al. 2013). Recent studies suggest that one factor is a type of neonicitinoid pesticide called imidacloprid, which has been shown to be deadly to bees (Chensheng et al. 2012) when ingested, even through plant sap droplets exuded from seeds once coated in neonicitinoids (Girolami 2009, Tapparo et al. 2011). On one level, this should be unsurprising since these chemicals are employed as pesticides that effectively kill insect pests; since bees are insects, it ought not be unexpected for them to die from it as well. Yet it is our blindness to the agency of the insects we employ—whether pollinating and honey-producing bees or mealworms that feed our iguanas, toads, and songbirds—and the material agency of the chemicals we deploy that keeps us

---

32 Do we become arboreal cyborgs because we have the solidified blood of a tree in our teeth? Ought we consider ourselves as such as a way to acknowledge the agency of the donor organism? Does this have implications for those who receive animal organ transplants?

from considering how something so basic as food presents different sorts of interactions to different types of creatures.

Unlike its siblings, the stronger problem is a more silent problem; it is more difficult to measure in conventional systems. But we all yearn to be stronger, even as we continue to disagree about what constitutes strength. But on a social level, we can probably agree: A stronger society is one that is sustainable and that is able to respond *effectively* to crises. A stronger system is one that does not just tolerate, but integrates difference and contradictions to become more useful. To achieve this stronger system, we need to cultivate the emergence of stronger agencies that invites others to participate and that celebrates the new voices that have become audible.

There is a tiny creature called a tardigrade, nicknamed "water bear" for its way of moving in the watery forests of moss, that has been tested in the most extreme of conditions. It has been dried out then successfully rehydrated.[33] It has been sent into the vacuum of space, only to revive unharmed upon its return. The tardigrade is a simple animal, made of few cells compared to larger life forms, and it does not do anything particularly well, other than survive. Yet its ability to not conquer, but to *persevere* through the harshest of conditions—to allow time to pass and then to come back to life—is a strength that few have and insufficient numbers value.

Perhaps we could all learn a lesson from this, the humble "water bear," who may in fact be the strongest of us all.

Conclusion: Expecting Thirdness

This book, the adventure that it took to you, the reader, was in the spirit of a hopeful message because it answers a looming yet unspoken hope from within and around the science studies and the philosophers of science: "The hope, in short, is that we might heed Wittgenstein's command to mind the differences between forms of thought and yet still hold onto the idea that we can have true beliefs about morality, or economics, or mathematics. We would have content diversity yet cognitive unity" (Lynch 2009: 50).

We began this venture with splurges of the wisdom of a seasoned mind, we entered the valley of the logoi of civilizations and rode against the phantasm of a civilization of the logos, continued with a dialog witnessed in silence, traveled from the cyberworld to the biomedicalization of the ethics of the womb, and listened to the signals that turn noise to understandings and lithium flowers: What, now, is it there is to learn from Restivo, the anarchist sociologist, Stingl, the daft punk, and Weiss, the modest cyborg?

It may be true that the kind of purification program such as "mechanical objectivity"—shown by Galison and Daston to have been specific to early modern

---

33   One specimen had been dehydrated and sealed in a container for decades with a piece of moss; it came back to life when it was rehydrated.

science or more precisely, its representations—would rarely be embraced by any scientific discipline in practice without reflection. However, it is between the branches of the sciences and scholarships, between sciences and their publics, between science and techno-scientific practices that such a "purity" card—the opposite of solidarity—is being played in the current inter-epistemic-cultural relations between sciences, scholarships, and their publics. The end of the "Humboldt system" in Germany, anti-science governance in the US, the shuttering[34] of humanities programs in the West, the West (still) versus the rest, science against intellectuals, techno-science against science, body versus mind and so on: these territories of dispute (or should we say "disputed territories"?) are merely symptoms of a techno-scientific *natureculture* that has become purified ... or should we say "utterly uncivilized"?

We have turned to a search for new horizons of the sphere of cultural techno-cultural integration, and what we found was that to reach new horizons, the old boundaries must be made explicit. And so that is what we have attempted to do in this volume. Boundaries such as the mathematical logos, boundaries between mathematical civilizations, boundaries between physiological and social cognition, boundaries between offline and online, boundaries of ethical justifications for the creation of new human life: these are boundaries that we have previously considered our horizon. Yet, *enabled by diffraction,* we discover that we have not yet even seen the true horizons behind the high border walls and fences we have erected between and in the form of disciplines.[35]

We have conceived of *ScienceCraft* as a mode of multi-*naturecultural* solidarity that we need to embrace as members in the *techno-sciencivic* sphere that connects the epistemic communities of stake- and stockholders in science, technology, and scholarship. Indeed, what Jeffrey Alexander points out for the *Civil Sphere* is equally valid for *techno-sciencivil* discourse, for its "institutions, organizations of communication and regulation":[36]

> [S]ocieties are not governed by power alone and are not fueled only by the pursuit of self-interest. Feelings for others matter, and they are structured by boundaries of solidarity. How solidarity is structured, how far it extends, what it is composed of—these are critical issues for every social order, and especially for orders that aim at the good life. (Alexander 2006: 3)

---

34 Or worse, the "businessification" or corporatization ...

35 We play here with the two-sidedness of the event, of the expression of the expressed form of discipline that Deleuze and Guattari play with in *A Thousand Plateaus.* Do we dare to seek to discover what lies behind the horizon of the event?

36 Strangely enough, Alexander (2006) does not emphasize the importance of matters like (techno-)science or biomedicine into his discussion of the Civil Sphere and its bounded non-civil spheres of "state, economy, religion, family, and community [... *which are*] fundamental to the quality of life and to the vitality of a plural order" (ibid.: 5).

We recognize that today's societies are formed and structured in the diffractive interdependency of the rhizomatically figured world of techno-scientific practices and governance. In light of this, we present the following challenge: what if these boundaries are stretched to the point where they dissipate into the horizons of what is actually possible to imagine fully?

In the qualification of boundaries and possibilities as "truly" dissipative and imaginable, we demarcate (a) the functionality of truth—it is a property that is inherent and inherent only in the objects and practices that are negotiated in techno-scientific epistemic communities—and (b) its pluralism—that truth, despite each truth component being a "single property," can "manifest in multiple ways" (Lynch 2006: 191ff.). Truth is, at the same time and because of temporality, "one and many"; in Deleuze's terms, because "pluralism = monism," we can conclude that truth is a univocity, and therein it is relationalist and intertwined in becoming with haecceity.

In understanding that truth (even the truth of mathematics must and can be accounted for in this manner), we can dissolve the Western colonial matrix and the white male privileged and imperial gazes that W.E.B. Dubois, Sandra Harding, and Walter Mignolo (among others) have made explicit, problematized, and critiqued. This matrix and its gazes have haunted the philosophies and histories of science studies in one guise or the other, be it logos, modest witness, or Justice Scalia. By accounting for the *standpoint* and *situatedness* of knowledge, the practice of *epistemic disobedience*, and *border thinking* through nomadic scholarship—meaning thinking in figurations of becoming and diffraction not in static, frozen moments—we can achieve the integration of science and scholarship and we can find the solidarity we need to genuinely develop a truthful and civic science. Whether this is a *de*colonial or *post*colonial option, whether this begins in science studies, philosophy, history, anthropology, or sociology is of less importance than the fact that it is achieved. And for science studies and the philosophy[37] of science studies, we agree with and expand Julian Go's position that this "does not entail only studying non-Western societies, postcolonial social formations, or imperialism and colonialism but rather insists upon an over-arching theoretical approach and ontology that emphasizes the interactional constitution of social units, processes, and practices across space" (forthcoming: 4).

---

37   That theory and philosophy are two names for the same thing has been recently stressed by Elizabeth Grosz (2011), and that theory and practice cannot be separated is clear before us, in paraphrasing Kant to craft a motto for *Nü-Pragmatism*: "Any theory without practice cannot perform nor be embodied, any practice without theory cannot achieve nor be enacted, there it follows that empowerment in and by practice cannot be without enablement in and by theory."

How to Govern?

This raises pragmatic concerns. For example, "How to govern?" is a question that not only fazes governments, politicians, and political parties, but also an increasing number of (world-)societal actors who are enmeshed in the making of science policy and politics of science itself—in particular in the spheres of *life itself*: biomedicine and environmentalism. These actors are subject to government while licensed to usurp governing positions over disbanded and unevenly organized societal collectivities, imbricated in regimes of governance that are post-legitimized, post-transparent, post-democratic procedures. Philosophers of science, STS researchers, and environmental historians together must critique the acceleration in the dependence on techno-scientific practices of knowledge and management beyond the State by making explicit the trans-nationality and historicity of the enmeshed trajectories "across spaces."

Any claim that pertains to legitimated representation, representative participation, or transparency enunciated by citizens of nation-states, extra-national conglomerates, and producers of techno-scientific knowledge is obtaining to its own historicity of stakes held in some form of statehood. Meanwhile, post-democratic society and techno-scientific embeddedness of decision-making processes on the trans-local scale[38] seem to suggest that people paradoxically proliferate and govern their existences in abandonment of the State. Concurrently, transnational corporate entities rematerialize as quasi-state entities because the embeddedness of claims, the political-economical institutions, and the social construction of individual political existences are the products of history that cannot be grasped by either history of science or political history in isolation. A comprehensive analytics of the convergent discourse and the exigencies of the historical realities of trans-nationalism between (biomedical) science and its publics requires a critique that is informed by Science and Technology Studies (STS).

The history of the post-democratic transformation of our understanding of the State and its relation with its interlocutors is inherently intertwined with the transnational history of the discourse of the regime of the biomedical sciences, the environmental institutions regime, the convergent emergence of regimes of techno-scientific governance, and their different publics.[39] *World polity* analyst John W. Meyer and *situational analyst* Adele C. Clarke have studied these histories from 1890 forward using different methods. Meyer et al. have argued that the global history and emergent institutional properties of the environmental regime can be accounted for by using either discourse analysis or institutional analysis, and opted for the latter. By contrast, Adele Clarke applied forms of discourse

---

38   Trans-local meaning, broadly, that stake-holders and stock-holders do not share the same space of causes and effects.

39   This viewpoint applies to all modern, Western epistemic regimes (Elzinga 1993: 130).

analysis to create an effective discursive understanding of the imbrications of biomedical knowledge in American society.

In "taking ideas and discourse seriously" (Schmidt 2010), a new analytic mode of *discursive institutionalism* has become available for historians who have not abandoned the *genealogical* ideal that understanding the history of events matters for changing them in an effort towards better (political/institutional) practices, and who accept warnings by STS historians (like Hans-Joerg Rheinberger) that recent shifts towards preferring micro-histories over deeper histories runs the danger of creating ineffective "mono-cultures of the mind" (Shiva 1996). Equipped with these tools, we argue that contained within the European and American histories of science and scholarship[40] and their publics in the nineteenth and early twentieth centuries is the birthplace of the historic *a priori* for the state of post-democracy and techno-scientific governance. These processes must be thoroughly uncovered and understood, not only to unlock the potentials of a genuine postcolonial science—as has been the most urgent demand within STS circles—nor to pay tribute to the historically normative task of "decolonizing Enlightenment" in the face of the recent issue of the "provincialization of Europe" (Chakrabarty 2007). In order to go beyond indulging in blind "creative destruction" to uncovering actual "creative differences," we must understand our own genealogy to create potentials for new practices. In other words, one can only create difference with others by comparing one's system to the differences—temporal, spatial, social—within that system. However, unlike Foucault proper or systems-theory, we suggest that the pragmatic form of *discursive institutionalism* employed (as one of several tools) offers a distinct advantage because it provides a genuine theory of statehood built on historical evidence from the transatlantic, transnational, transcultural discourse of the interlocked, translocal regimes of biomedicine and environmentalism.

On the basis of demands and suggestions for alternative forms of governance and novel conceptualizations of statehood: Does the techno-scientific state necessarily emerge from the governance practices of actors such as corporations, NGOs and collective actors formed by interested private citizens as stakeholders? Will it redefine the boundaries of empirical and theoretical concepts in displacing or misplacing the State? How does the state so placed *see* people, and how do the people *look* back at it? The new State theory is based on an analysis of the history of convergent discursive trajectories, a nomadic entity that gazes at us just as heteroscopically as we gaze back at it.

Postcolonial Science Studies

Surely, "after colonialism," "after orientalism," "after occidentalism," we have now reached a time in the sociological study of science and society that has been aptly named "postcolonial" and has found itself poised against "mono-cultures of the mind." And just as adequately have postcolonial scholars problematized the

---

40   That is, of scientific institutions and individual scientists and scholars.

dark side of bio-medical and techno-scientific commodification and governance. Our critique in this work is not intended to detract from these existing important and sophisticated contributions; instead we want to add an important argument that draws upon insights from the recent surge of a "New-" and "Vital Materialism."

We seek to enrich the current postcolonial science/society&political studies discourses on the conceptual politics of environment and nature by 1) problematizing tacit assumptions about agency and vital materiality of ecologies and 2) arguing that original contexts and repertoires of culture were formed around actual practices that were more attuned to accounting for these kinds of agencies. Contemporary political critique from the wide field of postcolonial studies tends to adopt the nineteenth century conceptual transformation from "circumstance to environment" uncritically with regard to the agencies at play. It also re-reads cultural contexts and repertoires (including their historic origins) without an account for these agencies and vital materialities, substituting the vital material ecological dialectics that exist between human and non-human agencies with a reduction to flawed binary self-/non-self pattern; this permeates biological and ecological thought at all levels.

Postcolonial science studies need not choose between adopting or critiquing the Western "science versus myth" or "high-tech versus low-tech" accounts of knowledge production and technological practice. We propose as a third alternative a re-conceptualization based on a complementary reading of Ron Eglash's analog/digital-distinction: While most discourses concerning the environment and wilderness problematize the digital dimension of ecological agency from the perspective of culture in a culture-nature divide (wherein Nature is the context for cultural repertoires to be formed), an analog account radically reconstructs the agencies at play as contiguous vital materialities and material vitalities that include cultural contexts and repertoires dialectically entangled in the play of agencies. We ought not abandon the digital dimension, for it serves an important purpose in bridging translocal expertise and lay audience communication. However, for research and policy action to be more than just "efficient governance," for it to be genuinely efficacious and widely *a*ffective (and not merely *e*ffective), it is impossible to avoid the complexities of these agencies. Moreover, original cultural contexts and repertoires can be reconstructed as having taken a more insightful analog account of these agencies represented in cultural practices as *animacies* (Chen 2012). When we juxtapose this concept with Postcolonial Science Studies, we see that *animacies* are already embodied and enacted in cultural contexts and repertoires as *silent Other(s)* in interactions between humans or humans/non-humans, which we can conceptualize in a (cultural) theory of agency and thirdness.

Thirdness

Thirdness is a key concept to bring into this discussion for three reasons. First, there is a notable turn from dyads to triads in modeling interactions and relationships that has been ongoing for decades but that recently has emerged into the collective consciousness, whether in Fleck's constructions of facts by scientific communities, Peirce's linguistics, or Serres' theories of multiplicity and *parasite*. This represents, from a larger perspective, the rejection of absolute truths and recognizes instead the importance of Haraway's "situated knowledges" to the creation of shared objectivity. Second, the emergence of a third entity, such as in the idea of the parasite (or noise), requires an acknowledgment of *process* in interactions; emergent vital materiality exists only as it moves through time. Third, the unexpected character of emergence, as with other complexity-based outputs like fractals, tends to represent unwanted incursions into sterile, planned systems. But by harnessing the uncertainty, by intentionally setting up an environment hospitable to producing emergent Thirds, we have the potential to find novel ways to engage in discourse that has been heavily oppositional. In this discussion, the Third can be (and simultaneously is) a third party observer, the iterated intra-actions between subjects, the emerging discourse itself—by granting any of these with a type of agency, we can engage them as something productive and creative. When the negotiations of boundaries of *thirded* agents fail, it is because limitations of embodiment apply. For the human topography, this results in intersections with the world that are renegotiated into orders of natural environments, social environments, and cultural environments. These relations within the human ecology [the *Within*] are dialectical, interactive, *wechselwirkend* interactions; we are becoming less *analog* and more *digital,* more diffracted, more *parallactic*, more displaced. And with that, our ecology turns the letter "N" in Nature into the more emphasized capital N. The nature humans as bodies actually really live (as furniture) in and the Nature that they furnish (and are garnish for)—Stingl and Weiss call this: *anthropocology*—are two diffracted things (it is the diffraction itself that is *anthropocology*). The human ontologies, as historic ontologies, are therefore figuratively (furnishment) and literally (garnishment) *problematic*. That is because they are referring to conceptualizations for surrounding ecologies that construct emerging and developing environments in relations that eventually form the human point of view to mean that there is no human *bios* without human *zoë* or human *ethos*.

In turn, science studies, medical humanities, critical thinking and natural philosophy are not disparate and segregated scholarly entities; they are actually as one, united in and by theory. Over the course of its history, from Al-Ghazali's *Tahāfut al-Falāsifa*[h] to recent bouts of anti-intellectualism and the war on scholarship and higher education, philosophy has been accused of being useless and impractical. The case of the human body illustrates how misplaced and foolish this accusation really is, inherently and consequently, because it is precisely the body that is at the same time subject and object of conceptual practices and the uses of words that obtain gravity of meaning and depth of consequence. Speaking through Ellen Langer, it does make a huge difference for a patient's life-course whether s/he has been "cured of cancer" or the "cancer has gone into remission." The context of this project is not to show that theory can have the occasional bout in touching upon the limits (*Grenze* [German], *frontière* or *confins* [French]) of relevance for practice (*praxis*); on the contrary, it is the example of the body as a dynamic, mobile, and intensive boundary (*Grenze, seuil* or *confins*) between the cultural entity of the "being human" and its nature that reveals that we cannot escape the comprehensive reality of philosophy—which is "theory's real name," according to Elisabeth Grosz.

If we understand the singular philosophical reality of the conceptualization behind the terms of the frontier, the limit, and the boundary with Donna Haraway as a *figuration*, we attain a conceptualization wherein sign and substance appear noumenally resolved as "material trope and tropic quality of materiality," as manifest tradition of human hopes in the inheritance of our times (*Erbschaft unserer Zeit*) that are meant to serve for a collective reflection. At the same time, possibilities of redemption and salvation, of damnation and failure, and of collective integration (of migration between discursive spheres) emerge. We attain a concept of the boundary which is at the same time a concept of the *threshhold* (*Schwelle, seuil*), and the theory/philosophy of the boundary is now an *encyclopedic study of the threshold,* a gesture in transgression of the threshold, a *geophilosophy* and productive *deterritorialization* that manifests in meanings that begin to constitute the quality of *bodiliness* itself. It is the body that moves across borders, that grows from threshold to threshold of maturity and ages, that lies between the limits of *psyche* as part of its culture and of *organism* as part of its nature that constantly negotiates the threshold experiences. The body exists as an anthropological "boundary object," as heterotopia, as multiplicity, as a medium of rationality, as a permanently styled and actualized malapropism of theory between the disciplines of philosophy, biomedicine, sociology, psychology, as well as an interdisciplinary practice (as problematization) oscillating in the narrative dialectics of techno-scientific seeing. The distinction and the rigorous demarcation and maintenance of the boundary as a confinement or border between nature and culture, between natural science and humanities, and between philosophy and science appears, if not already problematic in itself, at the very least ripe to be problematized. It is required and inescapable that now we begin the problematization of the concept of the human body. We accord primacy to the problematization of the body as *real*.

This philosophical perspective does not simply or mindlessly "step over" boundaries, but instead it makes explicit that all such *transgressions* are, simultaneously, experiences of the threshold, which do not represent paradoxes that are internal to philosophy but that have common points of reference within different philosophical circles. The political imagination is inherently bound to the many philosophical, medical, sociological and historical dimensions of *the body problematic*. In walking "a thin æstheo-ethic line," we seek to take these fundamental insights forward in developing them into *one* structural realism, semantic holism, and narrative empathy that allows for a concept of the body that can be experienced as the human condition of an *integrative medicine* and permanent experience (*Schwellenerfahrung*) of overcoming thresholds, in short, as an ethics of becoming: The "being human" is the entity that, in being together with others and their shared environments, is entangled in perpetually becoming.

All practices, the textual and the non-textual, the discursive and the non-discursive, are only meaningful in narrative, and only through meaning does agency become possible. All agency is therefore semantic, and all semantic is embedded in narratives. Regarding the self, it is in the form of a genuine somatic style, as part of personal development and creation and education of one's character and mind (*Gemüth* [Kant]) while embedded in a concrete (empirical) body culture, that we discover that only with an awareness of body-culture and genuine, psychosomatically rooted mindfulness, can we, framed in the philosophical anthropology of the human condition, construct an ethics of becoming that creates the conditions of possibility for us to become cyborgs or a companion species, for otherwise we would end up as Borg (or mechanical automatons).

In our final thought, we want to remind readers that this perspective is and must be informed by feminist theory of a kind. But it is not the feminist theory that is exclusive or comes from having been excluded. This is a feminist theory that is solidarity, perhaps even the true source of solidarity—*sciencivics*—and the achievement of the civil sphere.[41] *ScienceCraft* is a feminist theory, and feminist theory that achieves becoming is *ScienceCraft*.

---

41   This means to account for subjectivities including the subjectivities that are feminine, transsexual, transgender, etc., and also the subjectivities of maleness and masculinity. By invoking these subjectivities, we acknowledge that maleness was traditionally defined even as it defined femaleness as its (silenced) Other. We account for that by acknowledging without blaming but with allowing for stating *explicitly* values, that that is and the issue of history and how it has progressed historically … until now. Beyond these historical processes, constitutions, and experiences of being male and being its other, however, we can find, we can experience, we can think in, we can communicate in, and we can become genuine femininity.

Bibliography

Abend, G. "The Meaning of 'Theory.'" *Sociological Theory* 26, no. 2 (2008): 173–99.

Acemoglu, D. "Technical Change, Inequality, and the Labor Market." *Journal of Economic Literature* 40 (2002): 7–72.

Adams, V. "Randomized Controlled Crime: Postcolonial Science in Alternative Medicine Research." *Social Studies of Science* 32 (2002): 659–90.

Adkins, T. "Simmel and Simondon: From the Venture of Life to the Advent of Adventure." Last accessed on January 11, 2012 at: http://fractalontology. wordpress.com/2012/01/22/simmel-and-simondon-from-the-ventures-of-life-to-the-advent-of-adventure/.

Adorno, T. "Fortschritt (1962/1964)." *Gesammelte Schriften* (Stichworte. Kritische Modelle 2): 10, no. 2. Frankfurt aM: Suhrkamp, 2003.

Adorno, T., J. Habermas, H. Albert, et al. *The Positivist Dispute in German Sociology*. New York, NY: Harper Torchbooks, 1976.

Agamben, G. *Homo Sacer.* San Francisco, CA: Stanford University Press, 1998.

Ahmed, S. *On Being Included.* Durham, NC: Duke University Press, 2012.

Albert, H. *Between Social Science, Religion, and Politics. Essays on Critical Rationalism*, Amsterdam: Rodopi, 1999.

Aleksandrov, A.D., A.N. Kolmogorov and M.A. Lavrent'ev, (ed.). *Mathematics: Its Content, Methods, and Meanings.* Cambridge, MA: MIT Press, 1969 (orig. publ. in Russian in 1956).

Alexander, J. "Recent Sociological Theory between Agency and Social Structure." *Schweizerische Zeitschrift fuer Soziologie* 18 (1992): 7–11.

Alexander, J. "Theorizing the 'Modes of Incorporation': Assimilation, Hyphenation, and Multiculturalism as Varieties of Civil Participation." *Sociological Theory* 19 (2001a): 237–49.

Alexander, J. "The Long and Winding Road: Civil Repair of Intimate Justice." *Sociological Theory* 19 (2001b): 371–400.

Alexander, J. *The Civil Sphere.* Oxford: Oxford University Press, 2006.

Alexander, J. "Keeping Faith with *Civil Sphere* and my Critics." *Perspectives* 30, no. 1 (2007): 10–15.

Almond, G., M. Chodorow, and R.H. Pearce, *Progress and Its Discontents,* Berkeley, CA: University of California Press, 1985.

Ambrose, A., (ed.). *Wittgenstein's Lectures, 1932–35,* Blackwell, 1979.

Anderson, B. *Imagined Communities: Reflections on the Origin and Spread of Nationalism.* London and New York, NY: Verso, 1991.

Anderson, E. "The Epistemology of Democracy." *Episteme* 3 (2006): 9–23.

Andreasen, N.C. "DSM and the Death of Phenomenology in America: An example of unintended consequences." *Schizophrenia Bulletin* 33 (2007): 108–12.

Anscombe, G.E.M. *Intention*. Cambridge, MA: Harvard University Press, 1957.

Apel, K.O. *Understanding and Explanation*. Cambridge, MA: MIT Press, 1984.

Aquinas, T. *Summa Theologica*. 1920. Translated by Fathers of the English Dominican Province, 2008. Available online at: http://www.newadvent.org/summa/3169.htm (orig. published ca. 1273).

Aranyosi, I. "A New Argument for Mind-Brain Identity." *British Journal for the Philosophy of Science* 62 (2011): 489–517.

Argamon, S., M. Koppel, J. Fine, A. Shimoni. "Gender, Genre, and Writing Style in Formal Written Texts." *Text*, 23 no.3 (August 2003): 321–46.

Arkin, R. *Behavior-based Robotics*. Cambridge, MA: MIT Press, 1998.

Armstrong, D.M. *A Materialist Theory of the Mind*. London: Routledge & Kegan Paul, 1968.

Ascher, M. *Mathematics Elsewhere: An Exploration of Ideas Across Cultures,* Princeton, NJ: Princeton University Press, 2004.

Astington, J. "What is theoretical about the child's theory of mind?: A Vygotskian view of its development," in *Theories of Theories of Mind,* ed. Carruthers, P. and P.K. Smith (Cambridge: Cambridge University Press, 1996), 184–99.

Baber, Z. *The Science of Empire: Scientific Knowledges, Civilization, and Colonial Rule in India*. Albany, NY: State University of New York Press, 1996.

Bakhtin, M. *The Dialogic Imagination: Four Essays*. Austin, TX: University of Texas Press, 1981.

Bakhtin, M. *Speech Genres and Other Late Essays*. Austin, TX: University of Texas Press, 1986.

Bakker, H. "The Life World, Grief and Individual Uniqueness: 'Social Definition' in Dilthev, Windelband. Rickert. Weber, Simmel and Schutz." *Sociologische Gids* 42 no. 3 (1995): 187–212.

Bakker, H. "Historical Sociology and the Comparative Method: Re-examining Weber's 'Means of Coercion' as an Ideal Type Model." Paper prepared for presentation at the North Central Sociological Association (NCSA) Meetings in Cleveland, Ohio, April 1–3, 2004.

Bakker, H. "Weber, Elias, and the Semiotics of Civility: Pre-Modern, Modern and Post-Modern Capitalism and Trust." Paper prepared for Conference on Risk, Civility and Trust, Victoria College, University of Toronto, May 5–8, 2005.

Bal, M. *On Meaning-Making*. Sonoma, CA: Poleridge Press, 1994.

Bal, M. "Memory Acts: Performing Subjectivity." *Boijmansbulletin.nl* 1/2 (2001): 8–18.

Baldassarri, D. and M. Diani. "The Integrative Power of Civic Networks." *American Journal of Sociology* 113 (2007): 735–80.

Ball, W.W.R. *A Short History of Mathematics*. New York, NY: Dover Publications, 1960.

Ball, W.W.R. and E.T. Bell. *Men of Mathematics*, New York, NY: Simon and Schuster Touchstone Books, 1937/1986.

Barad, K. "Getting Real: Technoscientific Practices and the Materialization of Reality." *Differences* 10 no. 2 (1998): 87–128.

Barad, K. "Posthumanist Perfomativity." *Signs: Journal of Women in Culture and Society* 28 (2003): 801–31.

Barad, K. *Meeting the Universe Halfway: Quantum Physics and the Entanglement of Matter and Meaning.* Durham, NC: Duke University Press, 2007.

Bare, M. "Dissertation project: *Nurture over Nature and the Rediscovery of the Body in Parsons's General Theory of Action.*" Forthcoming dissertation, University of Chicago.

Barnard, C. *The Functions of the Executive.* Cambridge, MA: Harvard University Press, 1968 [1938].

Barrett, L.F. "Psychological construction: A Darwinian approach to the science of emotion." *Emotion Review* 5, no. 4 (2013): 379–89.

Barrett, L., P. Henzi, and D. Rendall. "Social brains, simple minds," in *Social Intelligence: From Brain to Culture,* ed. N. Emery, N. Clayton, and C. Frith (Oxford: Oxford University Press, 2008), 123–46.

Bartlett, T. "Angry Words." *Chronicle of Higher Education,* March 20, 2012. Last accessed July 1, 2013 at: http://chronicle.com/article/Researchers-Findings-in-the/131260/.

Basken, P. "NIH Pledges Action After Review Affirms Racial Gap in Grant Approvals." *The Chronicle of Higher Education,* August 18, 2011.

Bataille, G. *The Accursed Share.* New York, NY: Zone Books, 1991.

Beck, U. *Power in the Global Age.* Cambridge: Polity, 2006.

Beiser, F. *The Fate of Reason.* Cambridge, MA: Harvard University Press, 1987.

Bell, D. "Beware of False Prophets: Biology, Human Nature, and the Future of International Relations Theory." *International Affairs* 82 (2006): 479–96.

Bell, E.T. *Men of Mathematics.* New York, NY: Simon and Schuster, 1937.

Bellos, A. *Here's Looking at Euclid.* New York, The Free Press, 2010.

van Bendegem, J.P. *Finite, Empirical Mathematics, Outline of a Model.* Gent, Belgium: Rijksuniversiteit te Gent, 1987.

Bennett, J. *Vibrant Matter: A Political Ecology of Things.* Durham, NC: Duke University Press, 2010.

Bennett, T. and P. Joyce (eds). *Material Powers.* London: Routledge, 2010.

Benveniste, É. "*Civilisation. Contribution à l'histoire du mot*" (English: Civilisation. Contribution to the history of the word), 1954, published in *Problèmes de linguistique générale,* Editions Gallimard, 1966: 336–45 (translated by Mary Elizabeth Meek as *Problems In General Linguistics,* 2 vols, 1971).

Berlin, I. *Four Essays on Liberty,* London: Oxford University Press, 1969.

Bilefsky, D. "Sworn to virginity and living as men in Albania." *New York Times* June 23, 2008. Last accessed on November 20, 2012 at: http://www.nytimes.com/2008/06/23/world/europe/23iht-virgins.4.13927949.html?pagewanted = all.

Blanton, D. and L. Cook. *They Fought Like Demons: Women Soldiers in the Civil War.* New York, NY: Vintage, 2003.

Blizzard Entertainment. *Diablo*. Blizzard Entertainment, 1996.

Blizzard Entertainment. *World of Warcraft*. Vivendi Universal, 2004.

Block, N. "What is consciousness?" Last accessed on January 29, 2013 at: http://blogs.scientificamerican.com/observations/2013/01/28/what-is-consciousness-go-to-the-video/.

Bloor, D. *Knowledge and Social Imagery*. London: Routledge and Kegan Paul, 1976.

Blumenberg, H. *Die Legitimität der Neuzeit*. Frankfurt, aM: Suhrkamp, 1966.

Blumenberg, H. *Die Lesbarkeit der Welt*. Frankfurt aM: Suhrkamp, 1979.

Blumenberg, H. *Lebenszeit und Weltzeit*. Frankfurt aM: Suhrkamp, 1986.

Boesch, E. "The myth of lurking chaos," in *Between Biology and Culture,* ed. H. Keller, Y.H. Poortinga and A. Schölmerich (Cambridge: Cambridge University Press, 2002), 116–35.

Boos, B. and M. Niss (ed.). *Mathematics and the Real World*. Boston, MA: Birkhauser, 1979.

Bordogna, F. *William James at the Boundaries*. Chicago, IL: Chicago University Press, 2009.

Bostrom, N. "A History of Transhumanist Thought." *Journal of Evolution and Technology* 14, no.1, (2005a): 1–25.

Bostrom, N. "In Defense of Posthuman Dignity." *Bioethics* 19, no. 3 (2005b): 202–14.

Boutroux, P. *L'ideal scientifique mathematiciens dans l'antiquité et dans les temps modernes*. Paris: Presses Universitaire, 1919.

Bowker, G.C. and S.L. Star. *Sorting Things Out: Classification and Its Consequences*. Cambridge, MA: MIT Press, 1999.

Boyer, C. *A History of Mathematics*. New York, NY: Wiley, 1968.

Brandom, R. *Making it Explicit*. Cambridge, MA: Harvard University Press, 1998.

Brassier, R. *Alien Theory: The Decline of Materialism in the Name of Matter*, Doctoral Thesis, University of Warwick, 2001.

Brassier, R. "Axiomatic Heresy: The Non-Philosophy of Francois Laruelle." *Radical Philosophy* 121 (September/October 2003): 24–35.

Brassier, R. *Nihil Unbound: Enlightenment and Extinction*. London: Palgrave Macmillan, 2007.

Breazeal, C. *Designing Sociable Robots*. Cambridge, MA: MIT Press, 2002.

Breithaupt, F. *Kulturen der Empathie*. Frankfurt aM: Suhrkamp, 2008.

Breithaupt, F. "The Birth of Narrative from the Spirit of the Excuse. A Speculation." *Poetics Today* 32 (2011): 107–28.

Breithaupt, F. "A Three-Person Model of Empathy." *Emotion Review* 4 (2012a): 84–91.

Breithhaupt, F. *Kultur der Ausrede*. Berlin: Suhrkamp, 2012b.

Brewer, M.B. and L. Caporael. "An evolutionary perspective on social identity: revisiting groups," in *Evolution and Social Psychology*, ed. M. Schaller (Philadelphia, PA: Psychology Press, 2006), 143–61.

Brodie, J. "Reforming Social Justice in Neoliberal Times." *Studies in Social Justice* 1 (2007): 93–107.

Bronfenbrenner, U. "Toward an experimental ecology of human development." *American Psychologist* 32 (1977): 513–31.

Bronfenbrenner, U. *The Ecology of Human Development*. Cambridge, MA: Harvard University Press, 1979.

Brothers, L. *Friday's Footprint: How Society Shapes The Human Mind*. New York, NY: Oxford University Press, 1997.

Brothers, L. *Mistaken Identity: The Mind-Brain Problem Reconsidered*. State University of New York Series in Science, Technology, and Society. Albany, NY: State University of New York Press, 2001.

Bryant, L. "Of the Simulacra: Atomic Images (Lucretius)." April 24, 2012. Last accessed on January 18, 2013 at: http://larvalsubjects.wordpress. com/2012/04/24/of-the-simulacra-atomic-images-lucretius/.

Buck-Morss, S. *The Dialectics of Seeing: Walter Benjamin and the Arcades Project*. Cambridge, MA: MIT Press, 1989.

Buffy the Vampire Slayer. Created by Joss Whedon. WB/UPN. 1997–2003. Television Broadcast.

Burton, L. *Learning Mathematics: from Hierarchies to Networks*. London: Routledge, 1999.

Butler, J. *Bodies That Matter: On the Discursive Limits of "Sex."* New York, NY: Routledge, 1993.

Byrne, R.W. "Brute intellect," *The Sciences* 31, no. 3 (1991): 42–7.

Cajori, F. *A History of Mathematics*. New York, NY: Macmillan & Co., 1894.

Cajori, F. *A History of Mathematical Notations, I/II*. LaSalle, IL: 1928/1929.

Campbell, N.D. "Addressing 'the social' psychopharmacologically: towards a more social neuroscience of 'addiction." Invited talk. Conference: "Addiction, the brain, and society" at Emory University. Atlanta, GA. February 27, 2009.

Campbell, N.D. "Why can't they stop? A highly public misunderstanding of science." Invited talk. Conference on "anthropologies of addiction and subjectivity" at McGill University. Montreal, Quebec. April 24, 2009. To appear as chapter in *Addiction Trajectories,* ed. E. Raikel and W. Garriott (Durham, NC: Duke University Press, 2013).

Campbell, N.D "The spirit of St. Louis: addiction research and psychiatric epidemiology." Invited paper given at the conference on 'What is the history of psychiatric epidemiology?" Inserm. Paris, France. June 4, 2010a.

Campbell, N.D. "Situated pharmaco-logics: how 'addiction' became lodged in the brain." Invited paper given at the conference on 'Addiction(s): social or cerebral?" European Neuroscience and Society Network. Helsinki, Finland. September 8, 2010b.

Campbell, N.D. "Towards a critical neuroscience of 'addiction.'" *Biosocieties* 5, no. 1 (2010c): 89–104.

Campbell, Nancy D. *Gendering Addiction: The Politics of Drug Treatment in a Neurochemical World.* Basingstoke; New York, NY: Palgrave Macmillan, 2011.

Canguilhem, G. *The Normal and the Pathological.* New York, NY: Zone Books, 1991.

Canguilhem, G. *A Vital Rationalist.* New York, NY: Zone Books, 2000.

Cantor, M. *Vorlesungen ueber die Geschichte der Mathematik.* Leipzig, 1907.

Caporael, L., R.M. Dawes, I.M. Orbell, and A.J.C. van de Kragt. "Selfishness examined: cooperation in the absence of egoistic incentives." *Behavioral and Brain Sciences* 12 (1989): 683–739.

Caporael, L. "Foolish liaisons: the 'new' SSK/AI alliance" unpublished manuscript, STS department, Rensselaer Polytechnic Institute, Troy, NY. 1990.

Caporael, L. "Sociality: coordinating bodies, minds and groups." *Psycoloquy* 6, no. 1 (1995).

Caporael, L. and R.M. "Groups as the mind's natural environment," in *Evolutionary Social Psychology,* ed. J. Simpson, and D. Kenrick (Hillsdale: NJ: Lawrence Erlbaum, 1997), 317–43.

Caporael, L., D.S. Wilson, C. Hemelrijk, and K.M. Sheldon. "Small groups from an evolutionary perspective," in *Theories of Small Groups: Interdisciplinary Perspectives,* ed. M.S. Poole and A.B. Hollingshead (Thousand Oaks, CA: Sage, 2004), 369–96.

Caporael, L. "Evolutionary theory for social and cultural psychology," in *Social Psychology: Handbook of Basic Principles,* ed. E.T. Higgins and A. Kruglanski (New York, NY: Guildford, 2007), 3–18.

Cappelen, H. and J. Hawthorne. *Relativism and Monadic Truth.* Oxford: Oxford University Press, 2009.

Carhart-Harris, R.L., H.S. Mayberg, A.L. Malizia, and D. Nutt. "Mourning and melancholia revisited: correspondences between principles of Freudian metapsychology and empirical findings in neuropsychiatry." *Annals of General Psychiatry.* 7, no. 9 (2008).

Carroll, P. "Science, Power, Bodies: The mobilization of nature as state formation." *Journal of Historical Sociology* 9 (1996): 139–67.

Carroll, P. "Tools, instruments, and agents: Getting a handle on the specificity of engine science." *Social Studies of Science* 31 (2001): 593–625.

Carroll, P. "Material Designs." *Theory and Society* 31 (2002): 75–114.

Carroll, P. "Water and technoscientific state formation in California." *Social Studies of Science* 42 (2012): 489–516.

Chen, M.Y. *Animacies.* Durham, NC: Duke University Press, 2012.

Chensheng, L.U., K.M. Warchol, and R.A. Callahan. "In Situ Replication of Honey Bee Colony Collapse Disorder." *Bulletin of Insectology* 65, no. 1 (2012): 99–106.

Clark, A. *Being There.* Cambridge, MA: MIT Press, 1997.

Clark, A. "Whatever next? Predictive Brains, Situated Agents, and the Future of Cognitive Science." *Behavioral and Brain Sciences* 36, no. 3 (2010a): 181–204.

Clark, A. *Supersizing the Mind: Embodiment, Action, and Cognitive Extension.* New York, NY: Oxford University Press, 2010b.

Clark, W. *Academic Charisma and the Origins of the Research University.* Chicago, IL: Chicago University Press, 2006.

Clarke, B. and M.B.N. Hansen, ed. *Emergence and Embodiment: New Essays on Second-Order Systems Theory.* Durham, NC: Duke University Press, 2009.

Cohen, E. "The Placebo Disavowed: Or Unveiling the Bio-Medical Imagination" *Yale Journal for the Humanities in Medicine,* 2002 Last accessed on February 1, 2013 at: http://yjhm.yale.edu/essays/ecohenprint.htm.

Cole, M., B. Means. *Comparative Studies of How People Think.* Cambridge, MA: Harvard University Press, 1981.

Collins, R. "Toward a neo-Meadian sociology of mind." *Symbolic Interaction* 12 (1989): 1–32.

Collins, R. *Sociological Insight.* 2nd ed. New York, NY: Oxford University Press,1992.

Collins, R. *The Sociology of Philosophies.* Cambridge, MA: Belknap, 1998.

Collins, R. *Interaction Ritual Chains.* Princeton, NJ: Princeton University Press, 2005.

Collins, R. and S. Restivo. "Robber Barons and Politicians in Mathematics," *The Canadian Journal of Sociology* 8, no. 2 (1983): 199–227.

Connolly, W.E. *Neuropolitics: Thinking, Culture, Speed.* Minneapolis, MN: University of Minnesota Press, 2002.

Connolly, W.E. *Worlds of Becoming.* Durham, NC: Duke University Press, 2011.

Coole, D. and S. Frost, S. (ed.). *New Materialisms: Ontology, Agency and Politics.* Durham, NC: Duke University Press. 2010.

Corbyn, Z. "Black Applicants Less Likely to Win NIH Grants." *Nature* (August 18, 2011).

Courant, R., H. Robbins. *What is Mathematics?* New York, NY: Oxford University Press, 1996 (rev. by Ian Stewart, orig. publ. 1906).

Currie, G. "Narrative Representation of Causes." *Journal of Aesthetics and Art Criticism* 64, no. 3 (2006): 309–16.

Currie, G. "Framing narratives," in *Narrative and Understanding Persons,* ed. D. Hutto (Cambridge: Cambridge University Press, 2007), 17–42.

Currie, G. *Narratives and Narrators: A Philosophy of Stories.* Oxford: Oxford University Press, 2010.

Currie, G. "Empathy for objects," in *Empathy: Philosophical and Psychological Essays,* ed. P. Goldie and A. Coplan (Oxford: Oxford University Press, 2011), 82–95.

D'Ambrosio, U. *Ethnomathematics.* Rotterdam: Sense Publishers, 2006.

D'oro, G. "The gap is semantic not epistemological." *Ratio* 20 (2007): 168–78.

Daft Punk. *Harder, Better, Faster, Stronger.* CD. Virgin, 2001.

Daft Punk. *Electroma.* Film. Daft Arts/Wild Bunch, 2006.

Damasio, A. *Descartes' Error.* New York, NY: Grosset/Putnam, 1994.

Davidson, I. and W. Noble. "The archaeology of perception." *Current Anthropology* 39, no. 2 (1989): 125–55.

Davis, P.J. and R. Hersh. *The Mathematical Experience*. Boston, MA: Birkhaeuser, 1981.

Day, R. "History, Reason, and Hope: A Comparative Study of Kant, Hayek and Habermas." *Humanitas* 15 (2002): 1–24.

Decker, O., M. Weißmann, J. Kiess, and E. Brähler. *Rechtsextreme Einstellungen in Deutschland* 2010, Friedrich Ebert Stiftung, 2010.

Deleuze, G. and F. Guattari. *A Thousand Plateaus*. London: Continuum, 2004.

Dew, K. "Public Health and the cult of humanity: A neglected Durkheimian concept." *Sociology of Health and Illness* 29, no. 1 (2007): 100–114.

Dickerson, A. and F. Green. "The Growth and Valuation of Computing and Other Generic Skills." *Oxford Economic Papers* 56 (2004): 371–406.

DiMaggio, P. and B. Bonikowski. "Make Money Surfing the Web? The Impact of Internet Use on the Earnings of U.S. Workers." *American Sociological Review* 73, no. 2 (2008): 227–50.

Dolphijn, R. and I. van der Tuin. (ed.). *New Materialism*. Open Humanities Press, 2012.

Dorrie, H. *One Hundred Great Problems of Elementary Mathematics, Their History and Solution*. New York, NY: Dover, 1965.

Dostoevsky, F. "Notes from the Underground," in *The Best Short Stories of Dostoevsky*, (New York, NY: Random House, 2001), 95–213.

Douglas, M. *How Institutions Think*, Syracuse, NY: Syracuse University Press, 1988.

Drucker, P. *The Age of Discontinuity; Guidelines to Our Changing Society*. New York, NY: Harper and Row, 1969.

Drucker, P. *Post-Capitalist Society*. Oxford: Butterworth Heinemann, 1993.

Dunbar, R. *Grooming, Gossip, and the Evolution of Language*. Cambridge, MA: Harvard University Press, 1998.

Dunbar, R., C. Gamble, and J. Gowlett (eds). *Social Brain, Distributed Mind*. Oxford: Oxford University Press, 2010.

Dupre, J. *The Disorder of Things*. Cambridge, MA: Harvard University Press, 1997.

Durkheim, E. *The Elementary Forms of the Religious Life*. New York, NY: Collier Books 1961/1912.

Durkheim, E. *The Elementary Forms of Religious Life*. New York, NY: Free Press, 1912/1995 (trans. Karen Fields).

Dworkin, R. *Taking Rights Seriously*. Cambridge, MA: Harvard University Press, 1977.

Dwyer, J. *The Relationship Rights of Children*. New York, NY: Cambridge University Press, 2006.

Dwyer, J. "Parents' religion and children's welfare: debunking the doctrine of parents rights." *California Law Review* 82, no. 6 (1994): 1371–447.

Dwyer, J. and E. Vig. "Rethinking transplantation between siblings." *The Hastings Center Report* 25 (1995): 7–12.

Edwards, L. "Re-fashioning the warrior Hua Mulan: Changing norms of sexuality in China." *IIAS Newsletter*. 2008. 48: 6–7. Available online at: www.iias.nl/files/IIAS_NL48_0607.pdf.

Eglash, R. *African Fractals,* Piscataway, NJ: Rutgers University Press, 1999.

Eisenberg, E. "Karl Weick and the aesthetics of contingency." *Organization Studies* 27, no. 11 (2006): 1–15.

Elias, N. *Über den Prozess der Zivilisation.* 2 Vols. Frankfurt aM: Suhrkamp, 1976.

Elias, N. *Studien ueber die Deutschen. Machtkämpfe und Habitusentwicklung im 19. und 20. Jahrhundert.* Frankfurt aM: Suhrkamp, 1989.

Elias, N. *The Civilizing Process: The History of Manners—State Formation and Civilization,* trans. Edmund Jephcott. Oxford and Cambridge, MA: Basil Blackwell, 1994.

Elias, N. *The Civilizing Process: Sociogenetic and Psychogenetic Investigations.* Oxford: Basil Blackwell, 2000.

Ellul, J. *The Technological Society.* New York, NY: A. Knopf. 1964.

Elzinga, A. "The Continuation of Politics by Other Means.". in *Controversial Science: From Content to Contention,* ed. T. Rante, S. Fuller, and W. Lynch (Albany, NY: State University of New York Press, 1993), 127–51.

Emery, N., N. Clayton, and C. Frith. *Social Intelligence: From Brain to Culture.* Oxford: Oxford University Press, 2008.

van Englesdorp, Dennis, D.C. Foster, M. Frazier, N. Ostiguy, and J. Hayes. "Fall Dwindle Disease: A Preliminary Report." Pennsylvania State University, December 15, 2006.

Ernest, P. (ed.). *Mathematics, Education and Philosophy: an International Perspective.* London: The Falmer Press, 1994.

Ernest, P. *Social Constructionism as a Philosophy of Mathematics.* Albany, NY: State University of New York Press, 1999.

Esfeld, M. *Holism in Philosophy of Mind and Philosophy of Physics* (Synthese Library Volume 298). Dordrecht: Kluwer, 2001a.

Esfeld, M. "How a Social Theory of Meaning can be connected with Realism." *Facta Philosophica* 3 (2001b): 111–31.

Esfeld, M. "What are Social Practices?" *Indaga. Revista internacional de Ciencias Sociales y Humanas* 1 (2003): 19–43.

Esfeld, M. "The impact of science on metaphysics and its limits." *Abstracta* 2 (2006a): 86–101.

Esfeld, M. "La théorie de Bohm." *Sciences et avenir hors série. Le paradoxe du chat de Schrödinger* 148 (October/November 2006b): 71.

Esfeld, M. "Scientific realism and the history of science," in *The Controversial Relationships Between Science and Philosophy: a Critical Assessment,* ed. G. Auletta (Vatican City: Libreria Editrice Vaticana 2006c), 251–75.

Esfeld, M. "From being ontologically serious to serious ontology," in *Symposium on His Ontological Point of View,* ed. John Heil (Frankfurt, aM: Ontos-Verlag 2006d), 191–206.

Esfeld, M. "Metaphysics of science between metaphysics and science." *Grazer Philosophische Studien* 74 (2007): 199–213.

Esfeld, M. *Naturphilosophie als Metaphysik der Natur.* Frankfurt, aM: Suhrkamp, 2008b.

Esfeld, M. "The modal nature of structures in ontic structural realism." *International Studies in the Philosophy of Science* 23 (2009): 179–94.

Esfeld, M. "Physics and causation." *Foundations of Physics* 40 (2010b): 597–1610.

Esfeld, M. (ed.). *Philosophie der Physik*, Berlin: Suhrkamp, 2012.

Esfeld, M. "Ontic structural realism and the interpretation of quantum mechanics." *European Journal for the Philosophy of Science* 3 (2013): 19–32.

Esfeld, M., and Dorato, M. "GRW as an ontology of dispositions." *Studies in History and Philosophy of Modern Physics* 41 (2010a): 41–9.

Esfeld, M., and Lam, V. "Moderate structural realism about space-time." *Synthese* 160 (2008a): 27–46.

Esfeld, M. and V. Lam. "Holism and structural Realism," in *Worldviews, Science and Us. Studies of Analytical Metaphysics. A Selection of Topics from a Methodological Perspective*, ed. R. Vanderbeeken and B. D'Hooghe (Singapore: World Scientific 2010c), 10–31.

Fairlie, R.W. "Race and the Digital Divide." *Contributions to Economic Analysis & Policy* 3, no. 1 (2004). Available online at: http://www.bepress.com/bejeap/contributions/vol3/iss1/art15.

Ferreiros, J. "The Motives behind Cantor's Set Theory—Physical, Biological, and Philosophical Questions." *Science in Context* 17, no. 1/2 (2004): 49–83.

Flanagan, O. *The Really Hard Problem: Meaning in a Material World.* Cambridge, MA: MIT Press, 2007.

Fleck, L. *Genesis and Development of a Scientific Fact*, trans. Fred Bradley and Thaddeus J. Trenn. Chicago, IL: University of Chicago Press, 1979 (orig. published as *Entstehung und Entwicklung einer wissenschaftlichen Tatsache: Einführung in die Lehre vom Denkstil und Denkkollektiv.* Benno Schwabe & Co., Basel, Switzerland, 1935).

Fleming, J.W. *The Dark Side of the Enlightenment.* New York, NY: W.W. Norton, 2013.

Fodor, J. *The Modularity of Mind.* Cambridge, MA: MIT Press, 1983.

Fodor, J. and E. LePore. *Holism. A Shopper's Guide.* Oxford: Blackwell, 1992.

Forst, R. *Contexts of Justice. Political Philosophy beyond Liberalism and Communitarianism.* Berkeley, CA: California University Press, 2002 (orig. *Kontexte der Gerechtigkeit* Frankfurt aM: Suhrkamp, 2004[1994]).

Forst, R. "The Justification of Human Rights and the Basic Right to Justification. A Reflexive Approach." *Ethics* 120, 2010: 711–40.

Forst, R. *The Right to Justification. Elements of a Constructivist Theory of Justice.* New York, NY: Columbia Univeristy Press, 2011 (orig. *Das Recht auf Rechtfertigung.* Frankfurt aM: Suhrkamp, 2007).

Forst, R. *Toleration in Conflict.* Cambridge: Cambridge University Press, 2013 (orig. *Toleranz im Konflikt.* Frankfurt aM: Suhrkamp, 2003).

Forst, R. *Justification and Critique.* Cambridge: Polity, 2013 (orig. *Kritik der Rechtfertigungsverhaeltnisse.* Berlin: Suhrkamp, 2011).

Foucault, M. *Birth of the Clinic.* London: Routledge, 1989 (orig. *Naissance de la clinique.* Paris: Presses Universitaires de Frances, 1963).

Foucault, M. "Une histoire restée muette." *La Quinzenne littéraire* 8 (1966): 3f.

Foucault, M. "A preface to transgression." *Language, Counter-memory, Practice.* Ithaca, NY: Cornell University Press, 1977.

Foucault, M. *Power/Knowledge: Selected Interviews & Other Writings 1972–1977.* New York, NY: Pantheon Books, 1980.

Foucault, M. *Ethics: Subjectivity and Truth (Essential Works of Michel Foucault 1954–1984, 1),* ed. P. Rabinow, (New York, NY: The New Press), 1997a.

Foucault, M. *Ethics: Subjectivity and Truth, (Essential Works of Foucault—Volume One),* ed. P. Rabinow (New York, NY: Penguin), 1997b (1994).

Foucault, M. "The Masked Philosopher (1980)," in *Ethics: Subjectivity and Truth, (Essential Works of Foucault—Volume One),* ed. P. Rabinow (New York, NY: Penguin 1997c/1994), 321–9.

Foucault, M. *The Order of Things.* London: Routledge, 2002.

Fournier, M. *Marcel Mauss: A Biography.* Princeton, NJ: Princeton University Press, 2006.

Franklin, S. *Artificial Life.* Cambridge, MA: MIT Press, 1995.

Freire, P. *Education: the Practice of Freedom.* London: Writers and Readers Publishing, 1967.

Friedman, M. *A Parting of the Ways: Carnap, Cassirer and Heidegger.* Chicago, IL: Open Court, 2000.

Friedman, M. "Einstein, Kant and the A Priori." *Royal Institute of Philosophy Supplement* 63 (2008): 95–112.

Gallese, V. "Before and below 'theory of mind': embodied simulation and the neural correlates of social cognition." *Philosophical Transactions of the Royal Society, Biological Sciences* 362, no. 1480 (2007): 659–69.

Gamm, G. *Nicht Nichts.* Frankfurt, aM: Suhrkamp, 2000.

Garces, C. and A. Jones. "Mauss Redux: From Warfare's Human Toll to *L'homme total.*" *Anthropological Quarterly* 82, no. 1 (2009): 279–310.

Garfinkel, H. *Studies in Ethnomethodology.* Englewood Cliffs, NJ: Prentice-Hall, 1967.

Gasking, D. "Mathematics and the World," in *The World of Mathematics,* ed. J.R. Newman (New York, NY: Simon & Schuster, 1956), 1708–22.

Geertz, C. *The Interpretation of Cultures.* New York, NY: Basic Books, 1973.

Geertz, C. "Art as a Cultural System," in *Local Knowledge* ed. C. Geertz (New York, NY: Basic Books, 1983), 94–120.

"The Gender Genie." Last accessed 28 March, 2011 at: http://bookblog.net/gender/genie.php.

Gill, C. "Robopop" *Remix Magazine*, May 1, 2001.

Girolami, V., L. Mazzon, A. Squartini, et al. "Translocation of Neonicotinoid Insecticides From Coated Seeds to Seedling Guttation Drops: A Novel Way of Intoxication for Bees." *Journal of Economic Entomology* 102, no. 5 (October 1, 2009): 1808–15. doi:10.1603/029.102.0511.

Giroux, Henry A., "The new extremism and politics of distraction in the age of austerity." *Truth Out.* January 22, 2012. Last accessed on 23 January, 2013 at: http://truth-out.org/opinion/item/13998-the-new-extremism-and-politics-of-distraction-in-the-age-of-austerity.

Go, J. "For a postcolonial sociology." *Theory&Society* 42, no. 1 (2013): 25–55.

Goffman, E. *Frame Analysis.* New York, NY: Harper Torchbooks, 1974.

Goodman, N. *Fact, Fiction and Forecast.* Cambridge, MA: Harvard University Press, 1955.

Goodman, N. *Ways of Worldmaking.* Indianapolis, IN: Hackett, 1978.

Goody, J. *The Domestication of the Savage Mind.* Cambridge: Cambridge University Press, 1977.

Gopkin, S. "Theories and modules: creation myths, developmental realities, and Neurath's boat," in *Theories of Theories of Mind*, ed. P. Carruthers, P. and P.K. Smith (Cambridge: Cambridge University Press, 1996), 169–83.

Gordon, S. "Micro-sociological theories of emotion," in *Micro-sociological Theory: Perspectives in Sociological Theory*, ed. H.J. Helle, and S. Eisenstadt (Beverly Hills, CA: Sage, 1985), 133–47.

Gould, J.L. and C.G. Gould. *Animal Architects: Building and the Evolution of Intelligence.* Basic Books, 2007.

Greenblatt, S. *The Swerve: How the World Became Modern.* New York, NY: Norton, 2012.

Gregson, J. "System, environmental, and policy changes: Using the social-ecological model as a framework for evaluating nutrition education and social marketing programs with low-income audiences." *Journal of Nutrition Education* 33, no. 1 (2001): 4–15.

Grémaux, R. "Woman becomes man in the Balkans," in *Third Sex, Third Gender*, ed. G. Herdt (New York, NY: Zone Books, 1996), 241–84.

Greshoff, R. "Ohne Akteure geht es nicht!Oder: Warum die Fundamente der Luhmannschen Sozialtheorie nicht tragen." *Zeitschrift fuer Soziologie* 37 (2008): 450–69.

Grice, H.P. *Studies in the Way of Words.* Cambridge, MA: Harvard University Press, 1989.

Griffin, L.J. "Comment on *The Civil Sphere.*" *Perspectives* 30 (2007): 2–4.

Grosslight, J. "Cultivating a Discipline: Mersenne as a Mathematical Intelligencer." *History of Science,* forthcoming, manuscript provided by author.

Guerrero, L.K. and A.G. La Valley. "Conflict, emotion, and communication," in *The SAGE handbook of Conflict Communication*, ed. J.G. Oetzel and S. Ting-Toomey (Thousand Oaks, CA: Sage, 2006), 69–96.

Gumplowicz, L. *Grundrisse der Soziologie*. Vienna: Manz, 1905.

Gutting, G. *What Philosophers Know*. Cambridge: University of Cambridge Press, 2009.

Guyer, P. "Kant on the Systematicity of Nature: Two Puzzles." *History of Philosophy Quarterly* 20 (2003): 277–95.

Haas, P. "Do regimes matter? Epistemic communities and Mediterranean pollution control." *International Organization* 43 (1989): 377–403.

Haas, P. "Introduction: epistemic communities and international policy coordination." *International Organization* 46 (1992): 1–35.

Habermas, J. *Between Facts and Norms*. Cambridge, MA: MIT Press, 1998.

Hacking, I. "Natural Kinds," in *Perspectives on Quine*, ed. R.B. Barrett and R.F. Gibson, (Cambridge, MA: Blackwell, 1990), 129–41.

Haney, C., C. Banks, and P. Zimbardo. "Interpersonal dynamics in a simulated prison." *International Journal of Criminology and Penology* 1 (1973): 69–97.

Hanna, R. *Kant and the Foundations of analytical Philosophy*. Oxford: Oxford University Press, 2001.

Haraway, D. *Simians, Cyborgs and Women: The Reinvention of Nature*. New York, NY: Routledge, 1991.

Haraway, D. *Modest_Witness@Second_Millennium. FemaleMan_Meets_ OncoMouse*. New York, NY: Routledge, 1997.

Harding, S.. *Is Science Multicultural?* Bloomington, IN: Indiana University Press, 1998.

Harding, S.. *Sciences From Below*. Durham, NC: Duke University Press, 2008.

Harding, S., (ed.). *Postcolonial Science Studies Reader*. Durham, NC: Duke University Press, 2011.

Hart, K. "Marcel Mauss: In Pursuit of a Whole." *Comparative Studies in Society and History* 49, no. 2 (2007): 1–13.

Hatfield, E., J.T. Cacioppo, and R.L. Rapson. *Emotional Contagion*. Cambridge: Cambridge University Press, 1994.

Hawley, A.H. *Human Ecology: A Theory of Community Structure*. New York, NY: Ronald Press, 1950.

Hayles, N.K. *How We Became Posthuman: Virtual Bodies in Cybernetics, Literature, and Informatics*. Chicago, IL: University of Chicago Press, 1999.

Heelan, P. *Space-perception and the Philosophy of Science*. Berkeley, CA: California University Press, 1983.

Heidelberger, M. "Kantianism and Realism: Alois Riehl (and Moriz Schlick)," in *The Kantian Legacy in Nineteenth Century Science*, ed. M. Friedman and A. Nordmann (Cambridge, MA: Harvard University Press, 1992), 227–47.

Heinberg, R. *The End of Growth*. Gabriola Island, BC: New Society Pub., 2011.

Hengehold, L. *The Body Problematic*. University Park, PA: Pennsylvania State University Press, 2007.

Henrich, D. *Die Einheit der Wissenschafstlehre Max Webers*. Tübingen: J.C.B. Mohr, 1952.

Henrich, D. *Konstellationen*. Stuttgart: Klett-Cotta 1991.

Henrich, D. *Grundlegung aus dem Ich. Untersuchungen zur Vorgeschichte des Idealismus. Tübingen—Jena 1790–1794*. Frankfurt, aM: Suhrkamp, 2004.

Henrich, D. *Between Kant and Hegel: Lectures on German Idealism*, Cambridge, MA: Harvard University Press, 2008.

Hersh, R. *What is Mathematics, Really?* New York, NY; Oxford: Oxford University Press, 1999.

Hesse, M. *The Structure of Scientific Inference*. Berkeley, CA: California University Press, 1974.

Hintikka, J. "What is the axiomatic method?" *Synthese* 183 (2011): 69–85.

Hird, Myra J. *The Origins of Sociable Life: Evolution After Science Studies*. Basingstoke: Palgrave Macmillan, 2009.

Hoel, A.S., I. van der Tuin. "The ontological force of technicity: Reading Cassirer and Simondon diffractively." *Philosophy&Technology* 26, no. 2 (2013): 187–202.

Hogrebe, Wolfram. "Mantics and Hermeneutics." *Archiwum Historii Filozofii i Mysli Spolecznej (Archive of the History of Philosophy and Social Thought)*, Vol. 56 (2011): 233–45.

Hohfeld, W. *Fundamental Legal Conceptions as Applied to Judicial Reasoning*. New Haven, CT: Yale University Press, 1918.

Holland, D., W. Lachicotte Jr., D. Skinner, and C. Cain. *Identity and Agency in Cultural Worlds*. Cambridge, MA: Harvard University Press, 1998.

Hollis, M. *Models of Man. Philosophical Thoughts on Social Action*. Cambridge: Cambridge University Press, 1977.

Horton, R. *Patterns of Thought in Africa and the West: Essays on Magic, Science and Religion,* Cambridge: Cambridge University Press, 1997.

Howe, R. "Do parents have fundamental rights?" *Journal of Canadian Studies* 36, no. 3 (2001): 61–78.

Huemern, W. and C. Landerer. "Mathematics, experience and laboratories: Herbart's and Brentano's role in the rise of scientific psychology." *History of the Human Science* 23, no. 3 (2010): 72–94.

Huntington, S. *The Clash of Civilizations*. New York, NY: Simon & Schuster, 1996.

Iacoboni, M. *Mirroring People. The New Science of How We Connect*. New York, NY: Farrar, Straus & Giroux, 2008.

Ionnanidis, I. "Why most published research findings are false." *PLOS* 2 (2005): e124.

Israel, J. *Radical Enlightenment: Philosophy and the Making of Modernity 1650–1750*. Oxford: Oxford University Press, 2002.

Israel, J. *Enlightenment Contested*. Oxford: Oxford University Press, 2006.

Israel, J. *Democratic Enlightenment*. Oxford: Oxford University Press, 2011a.

Israel J. *A Revolution of the Mind: Radical Enlightenment and the Intellectual Origins of Modern Democracy.* Princeton, NJ: Princeton University Press, 2011b.

Jackson, T. *Prosperity without Growth.* London: EarthScan, 2009.

Jacobs, J. *The Death and Life of Great American Cities.* New York, NY: Vintage Books, 1992.

Jaspers, K. *Psychopathologie.* 1st edition. Berlin: Springer, 1913.

Jaspers, K. *Philosophie* Berlin: Springer, 1931–1932.

Jaspers, K. *Philosophie I-III.* 3 Vols. Berlin: Springer, 1932.

Jaspers, K. *Psychopathologie.* 4th edition. Berlin: Springer, 1946.

Jaspers, K. *Von der Wahrheit.* Muenchen: Piper, 1947.

Jaspers, K. *Der Arzt im technischen Zeitalter.* Muenchen: Piper, 1999.

Jennings, D. "Huggy Bear/Darlin'/Colm/Stereolab." *Melody Maker.* May 1, 1993, sec. Singles: Reviewed by Dave Jennings.

Jones, E. *Chaotic Justice.* Chapel Hill, NC: North Carolina University Press, 2009.

Joyce, P. *Democratic Subjects: The Self and the Social in Nineteenth Century England.* Cambridge: Cambridge University Press, 1994.

Joyce, P. *The Rule of Freedom.* London: Verso, 2003.

Joyce, P. "The potency of things: a critique of cultural history," in *Explaining Change In Cultural History*, ed. N. O'Ciosain (Dublin: University College Press), 2005.

Kamiyama, Kenji. *Ghost in the Shell: Stand Alone Complex.* DVD. Kokubunji, Tokyo, Japan: Production I.G., 2002.

Keen, S. "A Theory of Narrative Empathy." *Narrative* (Fall, 2006): 207–36.

Keen, S. "Strategic Empathizing: Techniques of Bounded, Ambassadorial, and Broadcast Narrative Empathy." *Deutsche Vierteljahrs Schrift* 82, no. 3 (Sept. 2008): 477–93.

Keen, S. "Psychological Approaches to Thomas Hardy," in *The Ashgate Research Companion to Thomas Hardy*, ed. R. Morgan (Farnham: Ashgate, 2010), 285–300.

Keen, S. "Empathetic Hardy: Bounded, Ambassadorial, and Broadcast Strategies of Narrative Empathy." *Poetics Today* 32, no. 2 (2011): 349–89.

Keen, S. "Narrative Empathy." *The Living Handbook of Narratology.* Hamburg: Hamburg University Press, 2012.

Keesing, R.M. "Models, 'folk' and 'cultural': paradigms regained?." *Cultural Models in Language and Thought*, ed. D. Holland and N. Quinn (Cambridge: Cambridge University Press, 1987): 369–93.

Kendall, L. "MUDer? I Hardly Know 'Er! Adventures of a Feminist MUDder," in *Wired Women: Gender and New Realities in Cyberspace*, ed. L. Cherny and E.R. Weise (Seattle, WA: Seal Press, 1996), 207–23.

Kenny, R.W. "Beyond the Elementary Forms of Moral Life: Reflexivity and Rationality in Durkheim's Theory" *Sociological Theory* 28 (2010): 215–44.

Kersting, W., (ed.). *Politische Philosophie des Sozialstaates.* Weilerwist: Velbrueck-Wissenschaft, 2000.

Kersting, W. *Kritik der Gleichheit.* Weilerwist: Velbreuck-Wissenschaft, 2005.

Kirksey, S.E. and S. Helmreich. "The Emergence of Multispecies Ethnography." *Cultural Anthropology* 25, no. 4 (2010): 545–76.

Kirsh, D. "Adapting the Environment Instead of Oneself." *Adaptive Behavior* 4, no. 3–4 (1996): 415–52.

Kirsh, D. "Distributed Cognition, Coordination, and Environmental Design" *Proceedings of the European Cognitive Science Society* (1999): 1–11.

Kirsh, D. and P. Maglio. "On distinguishing Epistemic from Pragmatic Actions." *Cognitive Science* 18 (1994): 513–49.

Kivisto, P. "Comment on *The Civil Sphere.*" *Perspectives* (2007): 5–7.

Klein, K.J., H. Tosi, and A.A. Cannella. "Multilevel theory building: Benefits, barriers, and new developments." *Academy of Management Review* 24 (1999): 243–8.

Kline, M. *Mathematics: A Cultural Approach.* Reading, MA: Addison-Wesley, 1962.

Knorr-Cetina, K.D. *Epistemic Cultures. How the Sciences Make Knowledge.* Cambridge, MA: Harvard University Press, 1999.

Knorr-Cetina, K.D. "Culture in global knowledge societies." *Interdisciplinary Science Reviews* 32 (2007): 361–75.

Kotov, K. "Semiosphere: A chemistry of being." *Sign Systems Studies* 30 (2002): 41–55.

Kramer, E.E. *The Nature and Growth of Modern Mathematics.* Princeton, NJ: Princeton University Press, 1970.

Kripke, S. *Wittgenstein on Rules and Private Language.* Cambridge, MA: Harvard University Press, 1980a.

Kripke, S. *Naming and Necessity.* Cambridge, MA: Harvard University Press 1980b.

Kroeber, A.L. and C. Kluckhohn. *Culture: A Critical Review of Concepts and Definitions,* Papers of the Peabody Museum of American Archaeology and Ethnology, Harvard University 47, no. 1. Cambridge, MA: Peabody Museum, 1952.

Kull, K. "Semiosphere and a dual ecology: Paradoxes of Communication." *Sign Systems Studies* 33 (2005): 175–89

Lance, M., J. O'Leary-Hawthorne. *The Grammar of Meaning.* Cambridge: Cambridge University Press, 1997.

Langings, J. *Conserving the Enlightenment: French Military Engineering from Vauban to the Revolution.* Cambridge, MA: MIT Press, 2004.

Laruelle, F. "A Summary of Non-Philosophy." *Pli: The Warwick Journal of Philosophy* 8 (1999): 138–48.

Latour, B., S. Woolgar. *Laboratory Life: The Social Construction of Scientific Facts.* Sage Library of Social Research. London: Sage Publications, 1979.

Latour, B. *We Have Never Been Modern,* trans. C. Porter. Cambridge, MA: Harvard University Press, 1993.

Latour, B. "How to talk about the body." *Body&Society* 10, no. 2/3 (2004): 205–29.

Latour, B. *Reassembling the Social*. Oxford: Oxford University Press, 2006.

Laycock, T. *Lectures on the Principles and Methods of Medical Observation and Research*. Philadelphia, PA: Blanchard and Lea, 1857.

Lenoir, T. *Instituting Science: The Cultural Production of Scientific Disciplines*. San Francisco, CA: Stanford University Press, 1997.

Lenoir, T. "The Goettingen School and the Development of Transcendental Naturphilosophie in the Romantic Era," *Studies in the History of Biology*, 5. Baltimore, MD: Johns Hopkins University Press, 1981: 111–205.

Lenoir, T. *The Strategy of Life*. Chicago, IL: Chicago University Press, 1982.

Lenski, G. *Human Societies*. New York, NY: McGraw-Hill, 1970.

Lerman, S. (ed.). *Cultural Perspectives on the Mathematics Classroom*. Dordrecht: Kluwer, 1994.

Lerner, M. "MRSA—Keeping up with the Neighbours." Last accessed on August 26, 2010 at: http://www.europeanhospital.com/en/article/7341-MRSA-Keeping_up_with_the_neighbours.html.

Levi-Strauss, C. *The Savage Mind*. Chicago, IL: Chicago University Press, 1966.

Levin, M. "Why we believe in other minds" *Philosophy and Phenomenological Research* 44 (1984): 343–59.

Levine, D. *Powers of the Mind*. Chicago, IL: Chicago University Press, 2008.

Levy-Bruhl, L. *How Natives Think*. Princeton, NJ: Princeton University Press, 1985/1926.

Lewin, K. *A Dynamic Theory of Personality*. New York, NY: McGraw-Hill, 1935.

Lindemann, G. "The Analysis of the Borders of the Social World: A Challenge for Sociological Theory." *Journal for the Theory of Social Behavior* l, no. 35 (2005): 69–98.

Lindemann, G. "Die Emergenzfunktion und die konstitutive Funktion des Dritten." *Zeitschrift fuerr Soziologie* 35 (2006): 82–101.

Lindemann, G. "Medicine as practice and culture. The analysis of border regimes and the necessity of a hermeneutics of physical bodies," in *Biomedicine as Culture: Instrumental Practices, Technoscientific Knowledge, and New Modes of Life*, ed. R.V. Burri and J. Dumit (New York, London: Routledge, 2007), 47–58.

Lindemann, G. *Das Soziale von seinen Grenzen her denken*. Weilerswist: Velbrück. 2009a.

Lindemann, G. "Gesellschaftliche Grenzregime und soziale Differenzierung." *Zeitschrift für Soziologie* 38 (2009b): 92–110.

Lindemann, G. "From Experimental Interaction to the Brain as the Epistemic Object of Neurobiology." *Human Studies* 32 (2009c): 153–81.

Lindemann, G. "Die Emergenzfunktion des Dritten – Ihre Bedeutung für die Analyse der Ordnung einer funktional differenzierten Gesellschaft," in *Zeitschrift für Soziologie* 39 (2010): 493–511.

Lindemann, G. "Neuronal Expressivity: On the Road to a New Naturalness," in *Neurocultures*, ed. F. Ortega and F. Vidal (Frankfurt/M., Berlin, Bern, Bruxelles, New York, Oxford, Wien: Lang, 2011a), 69–82.

Lindemann, G. *Das paradoxe Geschlecht. Transsexualität im Spannungsfeld von Körper. Leib und Gefühl.* Wiesbaden: VS Verlag für Sozialwissenschaften, 2011b.

Lindemann, G. "Die Kontingenz der Grenzen des Sozialen und die Notwendigkeit eines triadischen Kommunikationsbegriffs." *Berliner Journal fuer Soziologie* 22 (2012): 317–40.

Lindquist, K.A., T.D. Wager, H. Kober, E. Bliss-Moreau, and L.F. Barrett. "The brain basis of emotion: A meta-analytic review." *Behavioral and Brain Sciences* 35 (2012): 121–43.

Lindquist, K.A., E.H. Siegel, K. Quigley, and L.F. Barrett. "The hundred years emotion war: Are emotions natural kinds or psychological constructions? Comment on Lench, Flores, & Bench [2011]." *Psychological Bulletin* 139 (2013): 255–63.

Linklater, A. *Critical Theory and World Politics.* London: Routledge, 2007.

Lister, A. "The 'mirage' of social justice: Hayek against (and for) Rawls." *CSSJ Working Paper Series* SJ017 (2011): 1–30.

Lotman, J. "On the Semiosphere." *Sign Systems Studies* 33 (2005): 205–29.

Lotman, J. *Universe of the Mind: A Semiotic Theory of Culture.* London: I.B.Tauris, 1990.

Luhman, N. *Soziale Systeme.* Frankfurt, aM: Suhrkamp, 1984.

Luhmann, N. *Social Systems*, trans. J. Bednarz, Jr. Stanford: Stanford University Press, 1995.

Luhmann, N. *Die Gesellschaft der Gesellschaft.* Frankfurt, aM: Suhrkamp, 1997.

Lynch, M.P. *Truth as Many and One.* Oxford: Oxford University Press, 2009.

Lynn, J.A. *Giant of the Grand Siecle.* Cambridge: Cambridge University Press, 1997.

MacIver, R.M. *Community: A Sociological Study.* London: MacMillan, 1917.

Mackenzie, D. *Statistics in Great Britain 1865–1930,* Edinburgh: Edinburgh University Press, 1981.

Mahler, A. and M. Muslow. *Die Cambridge School der politischen Ideengeschichte.* Frankfurt aM: Suhrkamp, 2010.

Mann, M. "The Autonomous Power of the State: Its Origins, Mechanisms, and Results." *Archives européenes de sociologie;* 25, 1984: 185–213.

Mann, M. *Fascists.* Cambridge: Cambridge University Press, 2004.

Mann, M. *The Dark Side of Democracy.* Cambridge: Cambridge University Press, 2005.

Mannheim, K. *Ideology and Utopia.* New York, NY and Eugene, OR: Harcourt and Brace/Harvest Publishers, 1936.

Mao, W., M.A. Schuler, and M.R. Berenbaum. "Honey Constituents Up-regulate Detoxification and Immunity Genes in the Western Honey Bee Apis Mellifera." *Proceedings of the National Academy of Sciences* 110, no. 22 (April 29, 2013): 8842–6.

Marks, G., Miller, N. "Ten years of research on the false consensus effect." *Psychological Bulletin* 102 (1987): 72–90.

Marshall, A. *Principles of Economics.* London: Macmillan, 1890.

Maslow, A. "A theory of human motivation." *Psychological Review* 50 (1943): 370–96.

Mauss, M. "Rapports réels et pratiques de la psychologie et de la sociologie." *Journal de Psychologie Normale et Pathologique*, 21 (1924): 892–922 (German in: *Soziologie und Anthropologie* 2. Frankfurt aM: Fischer, 1989: 146–77).

Mauss, M. "A category of the human mind: the person." *Journal of the Royal Anthropological Institute* 68 (1938): 261–81 (German in: *Soziologie und Anthropologie* 2. Frankfurt aM: Fischer, 1989, 221–52; English in: Carrithers, M. et al, (eds), *The Category of the Person*. Cambridge: Cambridge University Press, 1985, 1–25).

May, C. "The Clinical Encounter and the Problem of Context." *Sociology* 41 (2007): 29–45.

May, C. and T. Finch. "Implementation, embedding, and integration: An outline of Normalization Process Theory." *Sociology* 43, no. 3 (2009): 535–54.

Mayo, E. *The Psychology of Pierre Janet*. London: Routledge & Kegan Paul, 1951.

McClain, E.G. *The Myth of Invariance: The Origins of the Gods, Mathematics, and Music from the Rig Veda to Plato*. Boulder, CO: Shambala, 1978.

McCulloch, W.S., Pitts, W. "A logical calculus of ideas immanent in nervous activity." *Bulletin of Mathematical Biophysics* 5 (1943): 115–33.

McDermid, D. "The Real World Regained? Searle's External Realism Examined." *Kriterion* 18 (2004): 1–9.

McGee, G. *Beyond Genetics*. New York, NY: Harper Collins, 2003.

McKnight, Steven A. "The Legitimacy of the Modern Age: The *Lowith-Blumenberg* Debate in Light of Recent Scholarship." *Political Science Reviewer* 27 (1998): 68–96.

McLeroy, K.R., D. Bibeau, A. Steckler, and K. Glanz. "An ecological perspective on health promotion programs." *Health Education Quarterly* 15 (1988): 351–77.

McRae, S. "Coming Apart at the Seams: Sex, Text and the Virtual Body," in *Wired Women: Gender and New Realities in Cyberspace*, ed. L. Cherny and E.R. Weise (Seattle, WA: Seal Press, 1996), 242–63.

McShea, D.W. "Metazoan Complexity and Evolution: Is There a Trend?" *Evolution* 50, no. 2 (1996): 477–92.

McTaggart, J.M.E. "The Unreality of Time." *Mind* 17 (1908): 457–74.

Merricks, T. *Truth and Ontology*. Oxford: Oxford University Press, 2007.

Merton, R.K. "Civilization and Culture." *Sociology and Social Research* 21 (1936): 103–13.

Merton, R.K. "The Matthew effect in science." *Science* 159, no. 3810 (1968a): 56–63.

Merton, R.K. *Social Theory and Social Structure*, enlarged ed., New York, NY: Free Press, 1968b.

Mesquita, M., S. Restivo, U. D'Ambrosio. *Asphalt Children and City Streets: A Life, a City, and a Case Study of History, Culture, and Ethnomathematics in São Paulo*. Rotterdam: Sense Publishers, 2011.

Metz, T. "An African Theory of Moral Status: A relational alternative to individualism and holism." *Ethical Theory and Moral Practice* 15 (2012): 387–402.

Metzloff, A.N. "The 'like me' framework for recognizing and becoming an intentional agent." *Acta Psychologica* 124 (2007): 26–43.

Meyer, J. *Weltkultur*. Frankfurt, aM: Suhrkamp, 2005.

Meyer, J., D.J. Frank, A. Hironaka, et al. "The Structuring of a World Environmental Regime, 1870–1990." *International Organization* 51, no. 4 (1997): 623–51 (German in Meyer [2005]: 235–74).

Mignolo, W.D. *Local Histories/Global Designs*. Princeton, NJ: Princeton University Press, 2000.

Mignolo, W.D. *The Darker Side of the Renaissance*. Ann Arbor, MI: University of Michigan Press, 2003.

Mignolo, W.D. "Epistemic Disobedience, Independent Thought, and De-Colonial Freedom." *Theory, Culture, and Society* 26, no. 7–8 (2009): 1–23

Mignolo, W.D. *The Darker Side of Western Modernity*. Durham, NC: Duke University Press, 2012.

Milgram, S. "Behavioral study of obedience." *Journal of Abnormal and Social Psychology* 67 (1963): 371–8.

Millar, W.R. "A Bakhtinian Reading of and Its Relationship to Social Space; Or, Finding the Lost Tribe of Levi and Why It Matters: A Study in Voice, Space, and Power," 2003. Last accessed on February 1, 2013 at: http://www.cwru.edu/affil/GAIR/papers/2003papers/millar.pdf.

Miller, D. *Principles of Social Justice*. Cambridge, MA: Harvard University Press, 2001.

Mills, C.W. "Language, logic, and culture," in *Power, Politics, and People* ed. I.L. Horowitz (New York, NY: Ballantine, 1963), 423–38.

Mithen, Steven. "Did farming arise from a misapplication of social intelligence?" in *Social Intelligence*, ed. N. Emery, N. Clayton, and C. Frith (Oxford: Oxford University Press, 2008), 353–74.

Moll, H. and M. Tomasello. "Cooperation and human cognition: the Vygotskian intelligence hypothesis," in *Social Intelligence*, ed. N. Emery, N. Clayton, and C. Frith (Oxford: Oxford University Press, 2008), 245–60.

Morris-Suzuki, T. *The Technological Transformation of Japan: From the Seventeenth to the Twenty-first Century*. Cambridge: Cambridge University Press, 1994.

Morrison, S.T. "A Hayekian theory of Social Justice." *NYU Journal of Law and Liberty* 1 (2005): 225–47.

Morton, T. *Ecology without Nature*. Cambridge, MA: Harvard University Press, 2007.

Morton, T. *Ecological Thought*. Cambridge, MA: Harvard University Press, 2010.

Mouffe, C. *On the Political*. London; New York, NY: Routledge, 2005.

Mouse Hunt. Directed by Gore Verbinski. Universal City, CA: DreamWorks Pictures, 1997. DVD.

Mukerji, C. "Unspoken Assumptions: Voice and Absolutism in the Court of Louis XIV." *Journal of Historical Sociology* 2 (1998): 283–315.

Mukerji, C. "Material Practices of Domination and Techniques of Western Power." *Theory and Society.* 31 (2002): 1–31.

Mukerji, C. "Intelligent Uses of Engineering and the Legitimacy of State Power." *Technology and Culture* 44 (2003): 655–76.

Mukerji, C. "Tacit Knowledge and Classical Technique in Seventeenth-Century France: Hydraulic Cement as Living Practice among Masons and Military Engineers." *Technology and Culture* 47 (2006): 213–33.

Mukerji, C. "The Territorial State as a Figured World of Power: Strategics, Logistics and Impersonal Rule." *Sociological Theory* 28, vol. 4 (2010): 402–42.

Mukerji, C. and B. Simon. "Out of the limelight." *Sociological Inquiry* 68, No. 2 (May 1998): 258–73.

Mulan. Directed by Tony Bancroft and Barry Cook. Los Angeles, CA: Walt Disney Pictures, 1998.

Murray, S. "Care and the Self: Biotechnology, Reproduction, and the Good Life." *Philosophy, Ethics, and Humanities in Medicine* 2 vol. 6 (2007). Available online at: http://www.biomedcentral.com/content/pdf/1747-5341-2-6.pdf

Muslow, M. and M. Stamm, (eds). *Konstellationsforschung.* Frankfurt aM: Suhrkamp, 2006.

Nasr, S.H., *Science and Civilization in Islam.* Chicago, IL: Kazi Publications, 2007.

Needham, J. *Science and Civilization in China. Volume III: Mathematics and the Sciences of the Heavens and the Earth.* Cambridge: Cambridge University Press, 1959.

Neugebauer, O. *The Exact Sciences in Antiquity.* Princeton, NJ: 1952[1969].

Nietzsche, F. *The Genealogy of Morals,* New York, NY: Doubleday-Anchor. 1956 [1887].

Nisbet, H.B. "Lucretius in 18th century Germany. With a commentary on Goethe's *Metamorphose der Tiere*." *Modern Language Review* 81 (1986): 97–115.

Noë, A. *Out of Our Heads: Why You Are Not Your Brain, and Other Lessons from the Biology of Consciousness.* New York, NY: Hill and Wang, 2010.

Nordman, A. "Object Lessons." *Scientiæ Studia* 10 (2012): 11–31.

Nordmann, A., B. Bensause, S. Loeve, et al. "Matters of Interest: The objects of research in science and techno-science," *Journal for General Philosophy of Science* 42 (2011): 365–83.

Norman, D. and D.E. Rumelhart. "Memory and knowledge," in *Explorations in Cognition,* ed. D. Norman and D. Rumelhart (San Francisco, CA: W.H. Freeman, 1975): 3–32.

Nozick, R. *Anarchy, State, and Utopia.* New York, NY: Basic Books, 1974.

Nye, A. *Words of Power: A Feminist Reading of the History of Logic.* New York, NY: Routledge, 1990.

O'Neill, O. *Constructions of Reason: Explorations of Kant's Practical Philosophy.* Cambridge: Cambridge University Press, 1990.

O'Neill, O. *Bounds of Justice.* Cambridge: Cambridge University Press, 2000.

O'Neill, O. *Autonomy and Trust in Bioethics*. The Gifford Lectures, University of Edinburgh, 2001.

O'Neill, O. and N. Manson. *Rethinking Informed Consent in Bioethics*. Cambridge: Cambridge University Press, 2007.

Oakes, G. *Weber and Rickert: Concept Formation in the Cultural Sciences,* Cambridge, MA: MIT Press, 1988.

Oetzel, J.G., S. Ting-Toomey, S. Rinderle. "Conflict communication in contexts: A social ecological perspective," in *The SAGE Handbook of Conflict Communication*, ed. J.G. Oetzel and S. Ting-Toomey (Thousand Oaks, CA: Sage, 2006).

Olson, D. "The cognitive consequences of literacy." *Canadian Psychology* 27 (1986): 109–21.

Ong, A. "Mutations in Citizenship." *Culture, Theory & Society* 23 (2006): 499–505.

Osserman, R. Review of János Bolyai, *Non-Euclidean Geometry, and the Nature of Space* (MIT Press, 2004) by Jeremy J. Gray; published in *Notes of the American Mathematical Society* 52, no. 9 (2005): 1030–34.

Parkhill, C. "Humanism after all? *Daft Punk's* existentialist critique of transhumanism." *Parrhesia* 8 (2009): 76–88.

Parsley, C. "The Mask and Agamben: the Transitional Juridical Technics of Legal Relation." *Law Text Culture Issue 1 Law's Theatrical Presence* 14 (2010).

Parsons, T. *The Social System*. New York, NY: The Free Press. 1951.

Perry, P. and A. Shotwell. "Relational Understanding and White Anti-Racist Practices." *Sociological Theory* 27 (2009): 33–50.

Pert, C. *Molecules of Emotion.* New York, NY: Scribners, 1997.

Pettit, P. "Consequentialism and Respect for Persons." *Ethics* 100 (1989): 16–26.

Pettit, P. *The Common Mind.* Oxford: Oxford University Press, 1996.

Pettit, P. "Responsibility Incorporated." *Ethics* 117 (2007): 171–201.

Pettit, P. "The Reality of Group Agents," in *Philosophy of the Social Sciences: Philosophical Theory and Scientific Practice*, ed. C. Mantzavinos (Cambridge: Cambridge University Press, 2009), 67–91.

Petryna, A. *Life Exposed: Biological Citizens after Chernobyl*. Princeton, NJ: Princeton University Press, 2002.

Picard, R. *Affective Computing.* Cambridge, MA: MIT Press, 1997.

Picoult, J. *My Sister's Keeper.* New York, NY: Atria Books, 2004.

Plotnitsky, A. "Berhard Riemann's Conceptual Mathematics and the Idea of Space." *Configurations* 17, no. 1 (2009): 105–30.

Pollilo, S. "The Network-Structure of the Self." Paper Presented at the ASA 2004, retrieved at allacademic.org on March 23rd, 2010.

Prado, G.C. *Searle and Foucault on Truth*. Cambridge: Cambridge University Press, 2006.

Prasad, A. "Scientific Culture in the 'Other' Theatre of 'Modern Science': An Analysis of the Culture of Magnetic Resonance Imaging Research in India." *Social Studies of Science* 3, no. 3 (2005): 463–89.

Prasad, A. "Beyond Modern versus Alternative Science Debate: Analysis of Magnetic Resonance Imaging Research." *The Economic and Political Weekly* 41, no. 3 (2006): 219–22.

Prasad, A. "Science in Motion: What Postcolonial Science Studies Can Offer." *Electronic Journal of Communication Information & Innovation in Health* (RECIIS) 2, no. 2 (2008): p. 35–47.

Priest, S. (ed.). *Theories of Mind,* New York, NY: Houghton Mifflin, 1992.

Quine, W.V.O. *From a Logical Point of View*. Cambridge, MA: Harvard University Press, 1953.

Quine, W.V.O. *Word and Object*. Cambridge, MA: MIT Press, 1960.

Rabinbach, A. *The Human Motor. Energy, Fatigue, and the Origins of Modernity.* Berkeley, CA: University of California Press, 1990.

Rabinow, P. *Essays on the Anthropology of Reason*. Princeton, NJ: Princeton University Press, 1996.

Radford, L. "Connecting Theories in mathematics education: challenges and possibilites." *ZDM Mathematics Education* 40 (2008): 317–27.

Rawls, J. *Theory of Justice,* Oxford: Oxford University Press, 1976.

Rawls, J. "Kantian Constructivism in Moral Theory." *Journal of Philosophy,* 77 (1980): 515–72.

Rawls, J. "Justice as Fairness: Political not Metaphysical." *Philosophy and Public Affairs* 14 (1985): 223–51.

Rawls, J. "The Idea of an Overlapping Consensus." *Oxford Journal of Legal Studies* 7 (1987): 1–25.

Rawls, J. *Justice as Fairness*. Cambridge: Harvard University Press, 2001.

Reed, I. "Justifying Sociological Knowledge: From Realism to Interpretation." *Sociological Theory* 26 (2008): 101–29.

Reich, W. "Three Problems of Intersubjectivity—and One Solution." *Sociological Theory* 28 (2010): 40–63.

Resnikoff, H.L. and R.O. Wells, Jr. *Mathematics and Civilization.* New York, NY: Dover, 2011.

Restivo, S. "Joseph Needham and the Comparative Sociology of Chinese and Modern Science: A Critical Perspective," in *Research in the Sociology of Knowledge, Sciences, and Art,* II, ed. R.A. Jones and H. Kuklick (Greenwich: JAI Press, 1979), 25–51.

Restivo, S. "Mathematics and the Limits of Sociology of Knowledge." *Social Science Information* 20, no. 4/5 (1981): 679–701.

Restivo, S. *The Social Relations of Physics, Mysticism, and Mathematics.* Dordrecht: Kluwer, 1983.

Restivo, S. *Mathematics in Society and History,* Dordrecht: Kluwer, 1992.

Restivo, S. "The Social Life of Mathematics," in *Math Worlds: Philosophical and Social Studies of Mathematics and Mathematics Education,* ed. S. Restivo, J.P. van Bendegem, and R. Fischer (Albany, NY: State University of New York Press, 1993), 247–78.

Restivo, S. *Science, Society, and Values: Toward a Sociology of Objectivity.* Bethlehem PA: Lehigh University Press, 1994.

Restivo, S. "Mathematics, mind, and society," in *Learning Mathematics* ed. L. Burton (London: Falmer Press, 1999), 119–34.

Restivo, S. *Mathematics in Society and History.* New York, NY: Kluwer, 2001/1992.

Restivo, S. "Bringing up and booting up: social theory and the emergence of socially intelligent robots." *Proceedings of the 2001 IEEE Conference on systems, man, and cybernetics,* Tucson, AZ, October 7–10, 2001.

Restivo, S. "Minds, machines, and bodies: the social turn," Plenary lecture, Luce faculty seminar on mind–computer interactions, the Claremont Colleges, Claremont, CA, March 7, 2003.

Restivo, S. "Einstein's brain, Napoleon's penis, and Galileo's finger: Toward a sociology of the brain," lecture in the Science Studies Program Colloquium series at the University of California at San Diego, San Diego, CA, November 24, 2003.

Restivo, S. "Of brains, robots, and minds: making sense of the 'society' in science, technology, and society," College of Liberal Arts at Rochester Institute of Technology, Rochester, New York, April 1, 2004.

Restivo, S. "Maths, minds, and machines: what does Einstein's brain mean, and did Shakespeare have one?" The honors academy at Brooklyn College City University of New York, Brooklyn, NY, May 10, 2004.

Restivo, S. "Theories of mind, social science, and mathematical practice," in *Perspectives on mathematical practices,* ed. J.P. Van Bendegem, B. van Kerkhove (Dordrecht: Kluwer, 2005), 61–79.

Restivo, S. "The travels of Einstein's brain, and did Shakespeare have one: Reading mind, body, and brain sociologically," Lecture at the Free University of Brussels, Brussels, Belgium, April 11, 2005.

Restivo, S. "Toward a sociology of the brain and some remarks on its applications," Lecture at the Learning Lab Denmark, University of Copenhagen, Copenhagen, Denmark, April 18, 2005.

Restivo, S. "The brain as a boundary object: theory and values at the intersection of the social and neurological sciences." Invited Lecture at Open Forum, California State Polytechnic Institute, Pomona, CA, May 5, 2005.

Restivo, S. "Mathematics," *Monza: Polimetrica. The Language of Science* (ISSN 1971–1352). June, 2007. (An overview of the sociology of mathematics written for an online encyclopedia.)

Restivo, S. "Ask Einstein's brain; and did Shakespeare have one? Reflections on the social life of the brain." Invited lecture, 27 November, James Martin Institute for Science and Civilization, Said Business School, Oxford University, Oxford UK, 2008.

Restivo, S. "Nicholas C. Mullins Lecture: The social life of the brain," Department of Science and Technology in Society, Virginia Technological University, Blacksburg, Virginia, April 24, 2009.

Restivo, S. *Red, Black, and Objective: Science, Sociology, and Anarchism.* Farnham: Ashgate, 2011.

Restivo, S. and W. Bauchspies. "The will to mathematics: minds, morals, and numbers," in *Foundations of science, 11, 1 & 2, special issue on mathematics: what does it all mean?* (ed. J.P. van Bendegem, B. Kerkhove, and S. Restivo) 11, no. 1–2 (2006): 197–215.

Restivo, S. and C. Beech. "Knowledge rituals: the social theory of mind, computing, and intelligence." Invited presentation, annual meeting of the American Sociological Association, Atlanta, GA, 2003.

Restivo, S. and R. Collins. "Mathematics and Civilization," *The Centennial Review* 26, no. 3 (1982): 277–301.

Restivo, S. and H. Karp. "Ecological Factors in the Emergence of Modern Science," in *Comparative Studies in Science and Society*, ed. S. Restivo and C.K. Vanderpool (Princeton, NC: Merrill, 1974), 123–43.

Restivo, S. and J. Loughlin. "The invention of science." *Cultural Dynamics* 12, no. 2 (2000): 135–49.

Reyna, S.P. *Connections: Brain, Mind and Culture in Social Anthropology,* New York, NY: Taylor and Francis, 2007.

Richards, R.J. *The Romantic Conception of Life.* Chicago, IL: Chicago University Press, 2002.

Ricoeur, P. "Universal Civilization and National Cultures," in *History and Truth*, ed. P. Ricoeur (Evanston, IL: Northwestern University Press, 1965), 271–84.

Rieger, S. *Die Individualität der Medien.* Frankfurt, aM: Suhrkamp, 2001.

Rieger, S. *Die Ästhetik des Menschen.* Frankfurt, aM: Suhrkamp, 2002.

Rifkin. J. *The Empathic Civilization.* New York, NY: Jeremy P. Tarcher/ Penguin, 2009.

Rifkin, J. "The Empathic Civilization," Address Before the British Royal Society for the Arts, 2010.

Rooney, D., G. Hearn, T. Mandeville, and R. Joseph. *Public Policy in Knowledge-Based Economies: Foundations and Frameworks.* Cheltenham: Elgar Publishing, 2003.

Rose, N. *The Politics of Life Itself.* Princeton, NJ: Princeton University Press, 2006.

Rosental, Claude. *Weaving Self-Evidence: A Sociology of Logic.* Princeton, NJ: Princeton University Press, 2008.

Rosenthal, G. *Erlebte und erzählte Lebensgeschichte. Gestalt und Struktur biographischer Selbstbeschreibungen.* Frankfurt aM: Campus, 1995.

Rosenthal, G. and A. Bogner, (eds). *Ethnicity, Belonging and Biography. Ethno-graphical and Biographical Perspectives.* Münster: LIT Verlag/New Brunswick: Transaction, 2009.

Rosenthal-Schneider, I. *Reality and Scientific Truth.* Detroit, MI: Wayne State University Press, 1980.

Roszak, T. *The Making of a Counter-Culture: Reflections on the Technocratic Society and its Youthful Opposition, with a new introduction.* Berkeley, CA: University of California Press, 1995 (orig. publ. 1969).

Rotman, B. *Ad Infinitum – The Ghost in Turing's Machine; Taking God out of Mathematics and Putting the Body Back In: An Essay in Corporeal Semiotics.* Stanford, CA: Stanford University Press, 1993.

Rousseau, D.M. and R.J. House. "Meso organizational behavior: Avoiding three fundamental biases," in *Trends in Organizational Behavior* ed. C.L. Cooper and D.M. Rousseau (New York, NY: John Wiley, 1994), 13–30.

Rowling, J.K. *Harry Potter and the Philosopher's Stone.* London: Bloomsbury. 1997.

Rowling, J.K. *Harry Potter and the Order of the Phoenix.* London: Bloomsbury, 2003.

Rubin, C. "Daedalus and Icarus revisited." *The New Atlantis* no. 8 (2005): 73–91.

Rundell, J. and S. Mennell, (ed.). *Classical Readings in Culture and Civilization.* London: Routledge, 1998.

Ryan, M.L. *Possible Worlds, Artificial Intelligence, and Narrative Theory.* Bloomington, IN: Indiana University Press, 1991.

Ryan, M.L. "Cyberage Narratology: Computers, Metaphor and Narrative," in *Narratologies: New Perspectives on Narrative Analysis*, ed. D. Herman (Columbus, OH: Ohio State University Press, 1999), 113–41.

Ryan, M.L. "Narrativity and its Modes as Culture-Transcending Analytical Categories." *Japan Forum* 21 (2011): 307–23.

Ryan, M.L. "Space," in *The Living Handbook of Narratology*. Available online: hup.sub.uni-hamburg.de/ihn/index.php/Space, 2012, last accessed on: February 1, 2013.

Sandel, M. *Liberalism and the Limits of Justice.* Cambridge: Cambridge University Press, 1982.

Sarasin, P. *Reizbare Maschinen: eine Geschichte des Körpers 1794–1914,* Frankfurt, aM: Suhrkamp. 2001.

Sassen, S. *Territory, Authority, Rights: From Medieval to Global Assemblages.* Princeton, NJ: Princeton University Press. 2006.

Scanlon, T. *What We Owe to Each Other.* Cambridge, MA: Harvard University Press (Belknap), 1998.

Schaefer, W. "Global civilization and local cultures: a crude look at the whole." *International Sociology*, 16 (2001): 301–19.

Schaerfe, H. "Possible Worlds in Narrative Space." *Impact, www.hum.auc.dk* (2002): 1–19.

Schmidgen, H. *Die Helmholtz-Kurven,* Berlin: Merve Verlag, 2010.

Schmidt, V. "Discursive Institutionalism: Scope, Dynamics, and Philosophical Underpinnings," in *The Argumentative Turn Revised: Public Policy as Communicative Practice*, F. Fischer and J. Forester, ed. Durham, NC: Duke University Press, 2012.

Schmidt, V.A. "Taking Ideas and Discourse Seriously: Explaining Change Through Discursive Institutionalism as the Fourth 'new Institutionalism.'" *European Political Science Review* 2, no. 1 (February 4, 2010): 1.

Schmieder, C. "World of Maskcraft vs. World of Queercraft?" *Journal of Gaming and Virtual Worlds* 1 (2009): 1.

Schmitt, P.P. "Der Tod an den Haenden" *FAZ* August 24, 2010. "Im mikrobiologischen Blindflug" *FAZ,* August 25, 2010. Retrieved August 27, 2010 at at www.faz.net.

Schoenrich, G. *Kategorien und Transzendentale Argumentation.* Frankfurt aM: Suhrkamp. 1981.

Schützeichel, R. "Methodologischer Individualismus, sozialer Holismus und holistischer Individualismus." *Das Mikro-Makro-Modell der soziologischen Erklärung: Zur Ontologie, Methodologie ind Metatheorie eines Forschungsprogramms.* Greve, J., Schnabel, A. Schützeichel, R, (eds). Wiesbaden: VS Verlag, 2009: 357–71.

Scott, J.C. *Seeing Like a State.* New Haven, CT: Yale University Press, 1998.

Searle, J.R. "Minds, brains, and programs." *Behavioral and Brain Sciences* 3, no. 3 (1980): 417–57.

Searle, J.R. *Expression and Meaning: Studies in the Theory of Speech Acts.* Cambridge: Cambridge University Press, 1985.

Searle, J.R. *Mind, Language, Society.* New York, NY: Basic, 1998.

Sedlacek, T. *Economics of Good and Evil.* New York, NY: Oxford University Press, 2011.

Seel, M. 2002. "Fuer einen Holismus ohne Ganzes" Seel, Martin, S*ich bestimmen lassen.* Frankfurt, aM: Suhrkamp, 2002: 89–101.

Sehon, S. "Evidence and Simplicity: Why we should reject homoepathy." *Journal of Evaluation in Clinical Practice* 16 (2010): 276–81.

Sehon, S. and D. Stanley. "A philosophical analysis of the evidence-based medicine debate." *BMC Health Services Research* 3 (2003): 14. Available online at: http://www.biomedcentral.com/1472–6963/3/14.

Selin, H. (ed.). *Mathematics Across Cultures.* New York, NY: Springer, 2001.

Serano, J. *Whipping Girl: A Transsexual Woman on Sexism and the Scapegoating of Femininity.* Jackson, TN: Avalon Publishing Group, 2007.

Serres, M. *Genesis,* trans. G. James and J. Nielson. Studies in Literature and Science. Ann Arbor, MI: University of Michigan Press, 1995.

Serres, M. *The Parasite,* trans. L. Schehr. Vol. 1. Posthumanities. Minneapolis, MN: University of Minnesota Press, 2007.

Sfard, A. "Mathematical practices, anomalies and classroom communication problems," in *Constructing Mathematical Knowledge,* ed. P. Ernest (New York, NY: Psychology Press, 1994), 248–73.

Shalin, D.N. "Signing in the Flesh: Towards a Pragmatist Hermeneutics." *Sociological Theory* 25 (2007): 193–224.

Shapin, S. "Discipline and Bounding: The History and Sociology of Science as Seen through the Externalism-Internalism Debate." *History of Science* 30 (1992): 333–69.

Shapin, S. and S. Schaffer. *Leviathan and the Air-Pump.* Princeton, NJ: Princeton University Press, 1985.

Sheng, A. and G. Xiao. "Micro-, Macro-, Meso-, and Metaeconomics." *project syndicate, Oct. 9, 2012*. Last accessed on October 11, 2012 at: http://www. project-syndicate.org/commentary/new-thinking-in-economics-by-andrew-sheng-and-geng-xiao.

Shiva, V. *Monocultures of the Mind*. London: Zed Books, 1993.

Shusterman, R. *Practicing Philosophy*. New York, NY: Routledge, 1997.

Shusterman, R. *Body Consciousness: A Philosophy of Mindfulness and Somaesthetics*. Cambridge: Cambridge University Press, 2008.

Simmel, G. "Der *Fragmentcharacter des Lebens*. Aus dem Vorstudien zu einer Metaphysik ([1916/17]." *Simmel, G. Gesamtausgabe 13* Frankfurt, aM: Suhrkamp, 2000: 202–16.

Sivin, N. "Introduction," in *Science and Civilisation in China: Volume 6, Biology and Biological Technology, Part 6, Medicine*, ed. J. Needham (Cambridge: Cambridge University Press, 2000), 1–37.

Skidelsky, E. "From epistemology to cultural criticism: Georg Simmel and Ernst Cassirer." *History of European Ideas* 29 (2003): 365–81.

Sloterdijk, P. "Atmospheric Politics," in *Making Things Public*, ed. B. Latour, B. and P. Weibel (Cambridge, MA: MIT Press, 2005), 944–53.

Smith, D.E. *History of Mathematics*. New York, NY: Dover, 1958.

Sohn-Rethel, A. "Science as Alienated Consciousness." *Radical Science Journal* no. 2/3 (1975): 65–101.

Sony Online Entertainment. *Everquest*. Sony Online Entertainment; 1999.

Sparacino, F. "Narrative Spaces," 2002. Last accessed February 1, 2013 at: http://www.generativeart.com/on/cic/papersGA2002/34.pdf.

Spengler, O. *The Decline of the West*. New York, NY: A. Knopf, 1926.

Stalnaker, Robert. *Inquiry*. Cambridge, MA: MIT Press, 1984.

Star, S.L. *Regions of the Brain: Brain Research and the Quest for Scientific Certainty*. Stanford, CA: Stanford University Press, 1989.

Starr, P. "Social Categories and Claims in the Liberal State." *Social Research* 59 (1992): 263–95.

Stehr, N. *Knowledge Societies*. London: Sage, 1994.

Stehr, N. (ed.). *The Governance of Knowledge*. New Brunswick, NJ: Transaction, 2003.

Stengers, I. and I. Prigogine. *Order Out of Chaos*. New York, NY: Bantam, 1984.

Stengers, I. *Power and Invention*. Minneapolis, MN: Minnesota University Press, 1997.

Stengers, I. *The Invention of Modern Science*. Minneapolis, MN: Minnesota University Press, 2000.

Stengers, I. "Thinking with Deleuze and Whitehead: A double test," in A. Cloots, K.A. Robinson, (eds), *Deleuze Whitehead and the Transformation of Metaphysics*. Brussels: KVAV. 2005a.

Stengers, I. "The Cosmopolitical Proposal," in *Making Things Public*, ed. B. Latour, and P. Weibel (Cambridge, MA: MIT Press, 2005b), 994–1003.

Sterelny, K. "Social intelligence, human intelligence and niche construction," in *Social Intelligence: From Brain to Culture*, ed. N. Emery, N. Clayton, and C. Frith (Oxford: Oxford University Press, 2008), 375–92.

Stichweh, R. *Zur Entstehung des modernen Systems wissenschaftlicher Disziplinen.* Frankfurt aM: Suhrkamp, 1984.

Stichweh, R. *Der frühmoderne Staat und die europäische Universität.* Frankfurt aM: Surhkamp, 1991.

Stingl, A. *Aufklaerung als Flaschenpost oder Anthropologie der Gegenwart.* Saarbruecken: VDM, 2009.

Stingl, A.I. "The ADHD regime and neuro-chemical selves in whole systems. A science studies perspective," in *Health and Environment*, ed. H. Kopnina, and H. Keune (New York, NY: Nova Science Publishing, 2010a), Chapter 7.

Stingl, A.I. "The Virtualization of Health and Illness." *TeloScope* 2010b. Available online: http://www.telospress.com/main/index.php?main_page = news_ article&article_id = 372.

Stingl, A.I. "Truth, Knowledge, Narratives of Selves. A Microclimatology of Truth." *American Sociologist* 42 (2011a): 207–19.

Stingl, A.I. "How to map the body's spaces." *Proceedings of the International Conference Myth-Making and Myth-Breaking,* Bucharest, 2011b.

Stingl, A. "Posthumanism," in *Impacts of Technological Change* (Sociological Reference Guide) Pasadena, CA: Salem, 2011c.

Stingl, Alexander I. 2012(n). Talks, available via academia.edu.
  a) "Semantic gaps, epistemic deficiencies and the cyborg gaze: Medical Imaging, gender, and the perspective of postcolonial philosophy of science" (NeuroGenderings II).
  b) "The Post-democratic body and the Bio-Scientific State" (ASA).
  c) "Styles of Suffering and Spaces of Pain: Somatic, Semantic, Narrative Sites of Empathy and Agency" (SSSI).
  d) "My body is dancing with a yodeling dog, the STS scholar said." (STS Italia).
  e) "Guerilla Paper: On nomadic Statehood."

Stingl, A.I. forthcoming. "The narrative dialectics technoscientific seeing."

Stingl, A.I. forthcoming. *Anthropos's Scaffoldings*. Contract under negotiation.

Stingl, A.I. "Review of Laura Hengehold *The Body Problematic*." *Foucault Studies* 1, no. 5 (2013).

Stingl, A., and Weiss, S. "Beyond and before the label: The ecologies and agencies of ADHD," in *Krankheitskonstruktionen und Krankheitstreiberei*, ed. M. Dellwing (VS Verlag, 2013).

Stingl, A.I., and S.M. Weiss. "Care Power information," Tentatively accepted. *Telos.*

Stokols, D. "Translating social ecological theory into guidelines for community health promotion." *American Journal of Health Promotion* 10 (1996): 282–98.

Stover, L. *The Cultural Ecology of Chinese Civilization.* New York, NY: Signet, 1974.

Struik, D. *A Concise History of Mathematics.* New York, NY: Dover Publications, 1967.

Struik, D. "The sociology of mathematics revisited: a personal note." S*cience and Society* 50 (1986): 280–99.

Sugimoto, M. and D.L. Swain. *Science & Culture in Traditional Japan.* Cambridge MA: MIT Press, 1978.

Sutherland, D. "The Role of Magnitude in Kant's Critical Philosophy." *Canadian Journal of Philosophy* 34, no. 3 (2004): 411–42.

Tapparo, A., C. Giorio, M. Marzaro, et al. "Rapid Analysis of Neonicotinoid Insecticides in Guttation Drops of Corn Seedlings Obtained from Coated Seeds." *Journal of Environmental Monitoring* 13, no. 6 (2011): 1564. doi:10.1039/c1em10085h.

Taylor, C. "What's wrong with negative liberty?," in *The Idea of Freedom*, ed. A. Ryan (Oxford: Oxford University Press, 1979), 175–93.

Taylor, C. *The Ethics of Authenticity.* Cambridge, MA: Harvard University Press, 1992.

Taylor, C. "Modernity and the Rise of the Public Sphere," in *The Tanner Lectures on Human Values 14* ed. G.B. Peterson (Salt Lake City, UT: University of Utah Press, 1993), 203–60.

Taylor, C. "Liberal Politics and the Public Sphere," in *New Communitarian Thinking: Persons, Virtues, Institutions, and Communities*, ed. A. Etzioni (Charlottesville, VA: University Press of Virginia, 1995), 183–221.

Taylor, C. "The Distance between the Citizen and the State," *Twenty-first Century*, 40: 1997.

Taylor, C. "Democratic Exclusion (and its Remedies?) The John Ambrose Stack Memorial Lecture," in *Citizenship, Diversity, and Pluralism: Canadian and Comparative Perspectives,* ed. A.C. Cairns, J.C. Courtney, P. MacKinnon, et al. (Montreal and Kingston: McGill-Queen's University Press, 1999), 265–8.

Taylor, C. *Wieviel Gemeinschaft braucht die Demokratie.* Frankfurt aM: Suhrkamp, 2002.

Thompson, E. *Mind in Life: Biology, Phenomenology, and the Sciences of Mind.* Cambridge, MA: Harvard University Press, 2008.

Tolkien, J. *The Return of the King.* London: George Allen & Unwin, 1955.

Torrance, S. "Real-world embedding and traditional AI," preprint, Middlesex University AI Group, UK, 1994.

Turner, V. *The Ritual Process: Structure and Anti-structure.* Chicago, IL: Aldine, 1969.

Tversky, B. "Narratives of Space time and Life." *Mind & Language* 19, no. 4 (September 2004): 380–92.

von Uexküll, J. *A Foray Into the Worlds of Animals and Humans With a Theory of Meaning*, trans. Joseph D. O'Neil. Minneapolis, MN: University of Minnesota Press, 2010 (orig. published as *Streifzüge durch die Umwelten von Tieren und Menschen.* Verlag von Julius Springer, 1934).

Valsiner, J. and R. van der Veer. *The Social Mind: Construction of an Idea.* Cambridge: Cambridge University Press, 2000.

Van Dijk, J.A.G.M. *The Deepening Divide: Inequality in the Information Society.* Thousand Oaks, CA: Sage, 2005.

Vandenberghe, F. "From Structuralism to Culturalism: Ernst Cassirer's Philosophy of Symbolic Forms." *European Journal of Social Theory* 4, no. 4 (2001): 479–97.

Varela, F.E. Thompson, E. and E. Rosch. *The Embodied Mind: Cognitive Science and Human Experience.* Cambridge, MA: MIT Press, 1991.

Verran, H. *Science and an African Logic.* Chicago, IL: The University of Chicago Press, 1992.

Vygotsky, L.S. *Mind in Society.* Cambridge, MA: Harvard University Press, 1978.

Vygotsky, L.S. *Thought and Language.* Cambridge, MA: MIT Press, 1986.

Walzer, M. *Spheres of Justice.* New York, NY: Basic Books, 1983.

Warschauer, M. *Technology and Social Inclusion: Rethinking the Digital Divide.* Cambridge, MA: MIT Press, 2003.

Weick, K. "Enacted Sensemaking in Crisis Situations." *Journal of Management Studies* 25 (1988): 305–17.

Weigel, S. *Body- and Image-Space.* London: Routledge, 1996.

Wertsch, J. *Voices of the Mind: A Sociocultural Approach to Mediated Action.* Cambridge, MA: Harvard University Press, 1991.

White. H. "The Narrativization of Real Events." *Critical Inquiry* 7 (1981): 793–8.

Whitehead, A.N. *Science and the Modern World.* New York, NY: Free Press, 1967 [1925].

Widmayer, S. "Schema Theory: An Introduction." Available online: http://www.mrjthompson.com/Documents/SchemaTheory.pdf. Last accessed November 20, 2012.

Wierzbicka, A. *Cross-Cultural Pragmatics, 2nd exp. ed.* Berlin: Mouton de Gruyter, 2003[1991].

Wiesing, U. "Immanuel Kant, his philosophy and medicine." *Medicine, Health Care and Philosophy* 11 (2008): 221–36.

Wigner, E., "On the Unreasonable Effectiveness of Mathematics in the Natural Sciences." *Communications on Pure and Applied Mathematics* 13 (1960): 1–14.

Wilkin, P. "Are you sitting comfortably? The political economy of the body." *Journal of Sociology of Health and Illness* 31 (2009): 35–50.

Wittgenstein. L. *Gesammelte Werke.* 8 Vols. Frankfurt, aM: Suhrkamp, 1885–1938.

Wittgenstein, L. *Philosophical Investigations.* New York, NY: MacMillan, 1953.

Woodhouse, B. *Hidden in Plain Sight.* Princeton, NJ: Princeton University Press, 2008: 15–47.

Wright, Ronald, *A Short History of Progress.* Philadelphia, PA: Da Capo Press, 2004.

Wrong, D.H. "The Oversocialized Conception of Man in Modern Sociology." *The American Sociological Review* 26 (1961): 183–93.

Yee, N. "The Norrathian Scrolls: A Study of EverQuest." Last accessed on March 10, 2011 at: http://www.nickyee.com/eqt/report.html, 2001.

Zahavi, D. *Subjectivity and Selfhood.* Cambridge, MA: MIT Press, 2008.

Zahavi, D. "Minimal Self and Narrative Self: A Distinction in Need of Refinement," in *The Embodied Self: Dimensions, Coherence and Disorders*, ed. T. Fuchs, H.C. Sattel and P. Henningsen (Stuttgart: Schattauer, 2010a), 3.

Zahavi, D. "Empathy, Embodiment and Interpersonal Understanding: From Lipps to Schutz." *Inquiry* 53 (2010b): 285–306.

Zajonc, R.B. "Feeling and thinking: preferences need no inferences." *American Psychologist* 35 (1980): 151–75.

Zajonc, R.B. "On the primacy of affect." *American Psychologist* 39 (1984): 151–75.

Zaslavsky, C. *Africa Counts: Number and Pattern in African Cultures*, 3rd ed. Chicago, IL: Lawrence Hill Books, 1999 (orig. publ. 1973 by Prindle, Weber, and Schmidt).

Zeleza, P.T. and I. Kakoma, I. *Science and Technology in Africa*. Trenton, NJ: Africa World Press, 2005.

Žižek, S. *Welcome to the Desert of the Real*. London: Verso, 2002.

Žižek, S. *Violence*. London: Picador, 2008.

Žižek, S. *First As Tragedy, Then As Farce*. London: Verso, 2009.

Index